JUPITER, THE LARGEST PLANET

JUPITER

THE LARGEST PLANET
ISAAC ASIMOV

Illustrated with photographs

Lothrop, Lee & Shepard Company ● New York

The author wishes to thank the following for permission to reproduce the photographs in this book:

Yerkes Observatory, University of Chicago, Williams Bay, Wisconsin—pages 14, 21, 30, 38, 47, 57, 78, 81, 85, 98, 100, 114, 130, 135, 151, 211

New York Public Library Picture Collection—page 32

Hale Observatories, California Institute of Technology—pages 34, 105, 132, 133, 160, 181

Lick Observatory, University of California, Santa Cruz, California—pages 61, 76, 80

Asimov, Isaac
   Jupiter, the largest planet.
   SUMMARY: A comprehensive study of Jupiter from the earliest discoveries of its distance, size, and satellites, to recent discoveries regarding its atmosphere, composition, and the Great Red Spot.
   1. Jupiter (Planet)—Juvenile literature. [1. Jupiter (Planet)] I. Title.
QB661.A84   523.4'5   72-9359
ISBN 0-688-40044-2
ISBN 0-688-50044-7 (lib. bdg.)

2   3   4   5   77   76   75   74

TO CHAUCY BENNETTS,
whose drive is entirely responsible

CONTENTS

LIST
OF
TABLES

Books on astronomy by
ISAAC ASIMOV

THE CLOCK WE LIVE ON
THE KINGDOM OF THE SUN
THE DOUBLE PLANET
ENVIRONMENTS OUT THERE
THE UNIVERSE
TO THE ENDS OF THE UNIVERSE
JUPITER, THE LARGEST PLANET

For young readers

THE MOON
MARS
STARS
GALAXIES
COMETS AND METEORS
THE SUN
ABC'S OF SPACE
WHAT MAKES THE SUN SHINE?

1

THE PLANETS

Naming the Planets

Even before the dawn of written history, people watching the night sky could not help noticing that a few stars were different from all the rest.

Most of the stars form patterns that stay the same night after night, year after year, lifetime after lifetime. They seem to be fixed to the curve of the sky and to turn with it, all in one piece. They are the "fixed stars."

A few of the brightest stars, however, do not stay in place compared to the other stars. They do not form part of a fixed pattern. One of these stars might be near a particular fainter star one night, a little farther away the next night, still farther the next night, and so on. Eventually such a star would make a complete circle of the sky, little by little. Going from west to east, it would finally return to the spot where it was first noticed.

Five such bright stars were noted by the early sky watchers.

The sun and moon also did not remain in the same spot with respect to the fixed stars. (Of course, we can't exactly see the sun against the background of stars. It is so bright that when it is in the sky, enough of its light scatters to turn the whole sky blue and to blot out anything else that may be present. However, the pattern of stars visible at night shifts slightly from night to night, and this is due to the motion of the sun against the stars. Because of this motion, different parts of the sky are blotted out and different parts remain visible.)

These heavenly bodies—the sun, the moon, and the five bright stars—were called *planetes* by the Greeks, because they wander among the stars. The word means "wanderers" and has come down to us as "planets."

The sun is the brightest object in the sky and the moon is next. They are the only objects in the heavens large enough to appear as circles of light. The moon makes up for its lesser brightness by having a shape that changes from night to night. It is sometimes a solid circle of light, sometimes a half circle, sometimes a thin crescent, sometimes anything in between.

The remaining five planets are not circles of light. They are bright dots like the stars—but brighter. The brightest of these five planets is sometimes, in fact, the brightest object in the heavens after the sun and the moon. It sometimes shines out in the western sky after the sun sets, the first of all the stars to appear as the sky darkens. It is then called "the evening star."

It doesn't always appear in the evening, though. Sometimes it shines late at night just before the sun is going to rise. As the sky grows lighter, it is the last of the stars to fade out. It is then called "the morning star."

In early times, people thought the evening star and the morning star were two different objects. Then they noticed that whenever the evening star was in the sky in the evening, the morning star was never present in the morning, and vice versa. It turned out that the two were a single planet which moved from one side of the sun to the other and back again. When it was on one side of the sun it was the evening star; when it was on the other, it was the morning star.

The evening star (or morning star) never moves very far from the sun. It is never visible later than three hours after sunset or earlier than three hours before sunrise. It is never visible at midnight. Never.

Another planet that is usually almost as bright as the evening star is quite different. It can be any distance from the sun and it can shine at all hours of the night. At midnight when the moon is not in the sky, this second-brightest planet would be the brightest object in the sky.

The people who watched the sky and studied the planets naturally wanted to give them names, so they could refer to them easily. The first people who studied the planets in detail were the Sumerians. They lived about five thousand years ago in the land we now call Iraq. They considered the planets sacred to their gods and gave each one the name of a god.

This habit is continued to this day. Even *we* call the planets by the names of gods.

The evening star, when it is shining at its brightest, is by far the most beautiful of the starlike planets, so it was named for the goddess of love and beauty. In biblical times the people of western Asia called it "Ishtar" for that reason.

The Greeks, in later centuries, got their first notions of astronomy from western Asia, and they adopted the same system of naming the planets. The evening star was called "Aphrodite," the name of their goddess of love and beauty.

Still later, the Romans borrowed the same scheme. They named the evening star "Venus," after *their* goddess of love and beauty. We still call it that today. The evening star is the planet Venus. The morning star is Venus also.

What about the second brightest of the planets? It isn't as bright as Venus, but it isn't the slave of the sun either. It can be visible in the sky at any time of the night. This greater freedom, combined with its great brightness, seemed to make it fit to be called by the name of the most powerful of the gods.

The people of western Asia called it "Marduk," the Greeks called it "Zeus," and the Romans called it "Jupiter." Again we follow the Roman way, and this second brightest of the planets is still called Jupiter by us today. It is the planet Jupiter that this book is about.

Venus and Jupiter are shown near each other in the sky. Magnification isn't high and each planet is just a blurry circle. It isn't hard to tell them apart, though. Jupiter is the one with the satellites.

We keep the Roman names for the other planets as well. For instance, there is a planet that, like Venus, shifts from side to side of the sun. It stays even closer to the sun than Venus, though, and shifts faster. It is Mercury, named for the messenger of the gods, who could move very speedily indeed.

Then there are two planets, beside Jupiter, that don't have to stay near the sun. One is distinctly reddish in color, so it is called Mars, after the Roman god of war. Finally, there is one that moves across the skies particularly slowly. It was as though it moved slowly because it was old, so it was named for an old god. The Greeks called it Cronos, naming it for the god who was the father of Zeus, and the Romans called it Saturn.

The Order of the Planets

Are the planets right there among the stars?

It seemed to the early astronomers that they couldn't be. The fixed stars seemed to be attached to the sky, all of them, so that they all turned in one piece. Since the planets moved from place to place among the stars, they surely could not be attached to the sky. They had to be somewhere between the starry sky and ourselves; they had to be nearer to us than the stars were.

What's more, it was quite possible that the various planets might be at various distances from us.

One way of deciding which planet might be closer to us and which might be farther is to consider the speed with which each moves from west to east across the starry patterns. The farther away a planet is, the bigger the circle it must make as it moves across the sky, and the longer it must take to complete that circle. If a faraway planet takes longer to make a complete circle of the sky than a nearby one does, that faraway planet must seem to move more slowly than the nearby one.

But how can we measure the speed of a planet's motion, the distance it travels in a particular time? We can't hold a yardstick up to the stars.

The ancient Sumerians worked out a system. They divided the circumferennce of any circle, including the large circle a planet makes in going completely around the sky, into 360 equal parts, which we call degrees. They chose that number partly because it is divided evenly by many smaller numbers, so that it is easy to use when fractions are necessary.

If any object moves completely around the sky, it is said to have moved 360 degrees, usually written 360°. If it moved halfway around the sky, say from the eastern horizon to the western horizon, it moved 180°. From the horizon to the zenith (the very highest point of the sky) is 90°, and so on.

Again, the width of the moon is such that if 700 moons were placed side by side in a straight line, they would stretch all around the sky. The width of the moon is therefore $^{360}\!/_{700}$ or just about half a degree, or 0.5° if you want to use decimals. The sun is about the same size as the moon in appearance and is also 0.5° wide.

If the moon is studied carefully from night to night, it turns out that it moves about 13° from west to east against the starry background in a single day. The sun, on the other hand, moves only about one degree in a day. (The fact that the sun moves completely around the sky in a little over 360 days—one degree a day—is another reason the Sumerians decided to divide circles into 360 equal parts.)

Since the moon travels thirteen times as fast as the sun does around the sky, it seems logical that the moon should be considerably closer to us than the sun is. We get proof of this because every once in a while the moon passes in front of the sun and produces an eclipse. It couldn't pass in front of the sun unless it were closer to us than the sun is.

The starlike planets have more complicated paths across the sky than the sun and moon do. Whereas the motion of both the sun and moon is quite steady, the starlike planets move at varying speeds. However, if we study them from night to night and take note of their *maximum* speeds, it turns out that Mercury and Venus can, at times, move faster than the sun, but never as fast as the moon. Mars, Jupiter, and Saturn each move more slowly than the sun at all times.

In the end, the Greeks decided that the moon was the closest of the planets and that beyond it came, in order, Mercury, Venus, the sun, Mars, Jupiter, and Saturn.

It was decided, then, that Jupiter, the planet of particular interest in this book, was the second most distant of the seven planets.

But how distant is the second most distant?

Nobody could tell in ancient times. For all that the very earliest stargazers could tell, all the planets were quite close. The sky didn't seem very high and it, along with everything in it, might just skim the tops of the highest mountains. After all, the clouds, which seem to be just under the sky, often gathered about the mountaintops. Maybe even the most distant planet was only a few dozen miles away.

By 300 B.C., however, the Greeks had become the outstanding astronomers in the world. They also developed geometry and learned to use geometric methods to calculate distances, so that they wouldn't have to measure them off with a yardstick.

About 230 B.C., for instance, a Greek astronomer named Eratosthenes measured the position of the sun in the sky from two different cities on the same day. He then used geometry to show that from the different positions in the sky and from the distance between the cities, the Earth must be a huge ball, or sphere, 8,000 miles from side to side (its diameter) and 25,000 around its middle (its circumference).

(It was hard to believe, at that time, that the Earth could possibly be so big, but Eratosthenes's calculation was proved to be correct when the time came, eighteen centuries later, that men sailed around the entire globe.)

About 130 B.C. another Greek astronomer, Hipparchus, used geometry to measure the distance of the moon. He found the distance of the moon to be about thirty times the diameter of the Earth, or about 240,000 miles. This also proved to be correct.

Once the moon's distance was known astronomers could tell, from the size it appeared to be, just how large it really was. It turned out that the moon was a sphere that was about 2,100 miles in diameter. It was not as large as the Earth by any means, but it was still big.

What about the other planets? The Greek astronomers knew that the sun was considerably farther from the Earth than the moon was. To appear as large as it seemed when it was so far away must mean that the sun was much bigger than the moon, and perhaps even bigger than the Earth. They could not tell exactly, though. They could not make observations in the sky that were accurate enough to judge the distance of any body but the moon, which was the nearest.

No Greek astronomer could possibly work out how far away Jupiter might be, except that it must be millions of miles away. Perhaps many astronomers of past centuries felt that mankind would never know, that the problem was just too difficult to work out.

The Orbits of the Planets

Yet, even so, people continued to study the sky. There were other problems that needed solving. For instance, there were puzzles in the way some planets moved among the stars.

The sun and the moon travel steadily from west to east among the stars. The sun makes its complete circle about the sky in 365¼ days. The moon, which moves a little over thirteen times as fast, makes its complete circle in 27⅓ days.

The other planets move in a more complicated way, however. Their speed varies, increasing at some times and slowing down at others. The outermost planets, Mars, Jupiter,, and Saturn, actually turn every once in a while and move backward, from east to west.

Jupiter, for instance, would move from west to east in usual fashion for about 39 weeks, then turn and move east to west for 17 weeks. Then it would go west to east again. The backward east-to-west motion is called "retrograde motion." While Jupiter makes its complete circuit of the sky in twelve years, it reverses its motion and moves retrograde twelve times, once each year.

This was a puzzle, and the system the Greeks made up to account for this peculiar motion of the outer planets was quite complicated.

It was not until nearly two thousand years after the Greeks had worked out a system by supposing the Earth to be the center of the universe (as it appeared to be) that someone decided to tackle the problem from a new standpoint altogether. The Polish astronomer Nicolaus Copernicus decided to try to work out a system where the sun would serve as the center of the universe.

By that time, after all, most astronomers were pretty certain that the sun must be considerably larger than the Earth, and very likely the largest object in the heavens. And it was certainly by far the brightest, so why shouldn't it be at the center?

In 1542 Copernicus published the details of his system, with the sun at the center and with all the planets moving around it, instead of around the Earth. In fact, he considered the Earth also to be moving around the sun.

The sun, by this system, couldn't be considered a planet because it was considered to be the motionless center of the planetary system. Around it were the planets in order.

First was Mercury, which circled the sun more closely than any other planet did. Second was Venus. Third was Earth (along with the moon, which still circled the Earth in Copernicus's system and therefore could no longer be considered a separate planet). Fourth was Mars, fifth was Jupiter, and sixth was Saturn.

The sun was thus circled by six planets, and all these bodies made up the "solar" system, from the Latin word for "sun" since it was the sun that was at the center. In the Copernican system, Jupiter was still the next to the farthest planet, but the distance was measured from the sun, and not from the Earth.

The Copernican system explained why the outer planets turned around and went east to west every once in a while. The Earth, traveling in a smaller circle than the planets farther from the sun did, moved more quickly. When the Earth was on the same side of the sun as Jupiter, for instance, Earth traveled more quickly and outstripped Jupiter. Jupiter seemed to fall behind and to be moving backward. (You see the same thing if two trains are on adjoining tracks moving in the same direction. If you are on the train that is going faster, you will see the other train seem to move backward.)

Then in 1609 a German astronomer, Johann Kepler, added another improvement. From very careful observations of the exact positions of the planet Mars in the sky from night to night, he showed that it did not move around the sun in an exact circle, as astronomers had till then believed. Instead, Mars's "orbit" (the path it took around the sun) was a curve called an ellipse, which looked like a somewhat flattened circle.

An ellipse has two "foci" (the singular is "focus"), one on either side of the center. The positions of the foci can be calculated geometrically for any ellipse.

The sun is always located at one focus of the ellipse followed by any planet going around it. This means that the planet Mars, for instance, is not always at the same distance from the sun. When it is on the same side of the ellipse as the focus containing the sun is, it is closer to the sun than when it is on the opposite side.

Kepler maintained that this was true of all the planets. They all travel around the sun in ellipses, some of which were more flattened than others. Every planet has a point in its orbit where it is closest to the sun. That is its "perihelion." At the opposite side of the orbit is a point where it is farthest from the sun. That is its "aphelion."

Kepler was able to work out the exact shapes of all the orbits. He found that Venus's orbit is an ellipse that is so slightly flattened that it is almost a circle. Mars's orbit is considerably more flattened. Astronomers don't like to say "slightly flattened" and "more flattened," however. Whenever possible, they like to measure something exactly and describe it with a number.

For instance, the flatter an ellipse is, the greater the distance between the two foci. The distance becomes a larger and larger fraction of the total distance across the long diameter

Johann Kepler was born in southwestern Germany in 1571. In 1609 he suggested that planets move around the sun in ellipses. This made it possible to work out the actual structure of the solar system for the first time. He was the first to suggest the use of the word "satellites" for worlds that circle planets. He died in 1630.

(the "major axis") of the ellipse as the ellipse becomes more flattened. This fraction is called the "eccentricity" of the ellipse.

For a circle, the two foci are located exactly at the center. There is zero distance between the foci, and the fraction of the diameter of the circle represented by the distance between the foci is also zero. For that reason we say that the eccentricity of a circle is equal to o.

If you take a *very* flattened ellipse, shaped something like a long thin cigar, the foci are almost at the opposite ends. The distance between the foci is almost equal to the major axis itself. The eccentricity of a very flattened ellipse is therefore close to 1.

When Kepler determined the orbits of the planets around the sun, he found the ellipses were not very flattened. The eccentricities of the orbits of each of the six planets he knew about is given in Table 1.

(This book will include a number of tables. It is always much more useful to give exact figures than to use inexact words. The actual statistics will give you a clearer picture of Jupiter than mere words would.)

As you see in Table 1, the eccentricities (in all cases but that of Mercury) are quite small, which means that the ellipses are fairly close to circles. The difference between perihelion and aphelion is small in such cases, and the sun is not much closer to planets with small eccentricities (such as the Earth) at one time than at another. That is why the sun always looks just about the same size to us.

The eccentricity of Jupiter's orbit is a little greater than Earth's, but not much more so. (In those tables where figures are given for different planets, the entry for Jupiter is in bold-face type so that you can see it at once.)

Another thing one could determine about the orbits when they are worked out by Kepler's system is that all are in almost the same plane. That is, suppose you imagine a huge flat sheet of paper spreading out from the sun in all directions in such a

TABLE 1

Eccentricities of Orbits

PLANET	ECCENTRICITY
Mercury	0.206
Venus	0.007
Earth	0.017
Mars	0.093
Jupiter	**0.048**
Saturn	0.056

way that the Earth moves around the sun exactly along the stretch of the paper. That sheet of paper would represent the plane of Earth's orbit and that plane is called the "ecliptic."

It turned out that all the other planets move along orbits that were very close to that plane. Each orbit would be tilted slightly to the ecliptic, but not much.

This tilt, or the "inclination" of the orbit, is measured as an angle of so many degrees.

To see what this means, imagine two lines meeting at one end. One of the lines is fixed, while the other turns. If the turning line makes a complete circle, it has moved through 360°. If it starts at the fixed line, makes a quarter turn and stops, the two lines make an angle of 90°. If the turning line starts at the fixed line and moves $\frac{1}{360}$ of the way around the circle, the two lines make an angle of 1°. A turn of only $\frac{1}{360}$ of the way around the circle isn't much, so an angle of 1° leaves the two lines quite close together.

In Table 2 we have the inclinations for each of the planets known to Kepler. These angles are measured from the ecliptic, the plane in which the Earth rotates. (Earth's inclination is therefore 0°.) Again except for Mercury, the tilt isn't much. Jupiter's inclination is only 1.3°.

As you see, then, as far as eccentricity and inclination are concerned, Jupiter is not particularly different from the other planets, or remarkable in any way. In fact, if any planet seems notably out of step, it is Mercury.

But let us go on—

TABLE 2

Inclinations of Orbits

PLANET	INCLINATION (IN DEGREES)
Mercury	7.0
Venus	3.4
Earth	—
Mars	1.9
Jupiter	**1.3**
Saturn	2.5

The Brightness of the Planets

Kepler's careful model of the solar system explained the reason why the brightness of the planets varied from one time to another.

Consider Mars, for instance. As we work our way outward from the sun, Mars is the first planet we come to beyond Earth. When Earth and Mars are on the same side of the sun, they can be quite close together. By Kepler's model, Mars is sometimes only a third as far from the Earth as the sun is.

On the other hand, when Mars and Earth are on opposite sides of the sun, then the two are separated by a far greater distance. The farthest separation of Mars and Earth, when the two are on separate sides of the sun, is seven times as great as when the two are on the same side of the sun.

This should affect the brightness of Mars. Naturally, when Mars is close to the Earth it should appear much brighter than when it is far away. It should, in fact, appear fifty times as bright when it is closest to us as when it is farthest from us.

And it does!

This was one of the reasons that the idea of a sun-centered planetary system made more sense than the old Greek notion of an Earth-centered planetary system. The Greeks had a hard time explaining why Mars was so much brighter at one time than at another. With the new scheme of Copernicus and Kepler, the reason became obvious.

But how do we express brightness in numbers?

Hipparchus, in ancient Greek times, called the brightest stars "first magnitude," while the very dimmest ones were of the "sixth magnitude." In between he placed stars of the second, third, fourth, and fifth magnitude. He made his judgments by eye.

In modern times, instruments were invented that measured the brightness of stars very accurately, far better than Hipparchus could manage with his eyes alone. It turned out that the dimmest stars were only about $\frac{1}{100}$ as bright as the brightest ones.

Astronomers decided to make each magnitude 2.5 times as bright as the magnitude below. A star of magnitude 1 was 2.5 times as bright as a star of magnitude 2; which was, in turn, 2.5 times as bright as magnitude 3, and so on. In that way a star of magnitude 1 was $2.5 \times 2.5 \times 2.5 \times 2.5 \times 2.5 = 100$ times as bright as a star of magnitude 6.

Astronomers could measure the brightness of stars so well that they could say a particular star was of magnitude 2.4 or 3.5.

A magnitude of 1.0 was set as the average of the twenty brightest stars. That means that half of them were brighter than average and had magnitudes of less than 1. The star Alpha Centauri has a magnitude of 0.3. The brightest star in the heavens, Sirius, has a magnitude that is less than zero on this scale and extends into the negative numbers. It is —1.4.

(In dealing with magnitude measurements, always remember that the smaller the magnitude the brighter the star, and that negative magnitudes are particularly bright.)

In Table 3, the magnitudes of each of the planets known to Kepler are given for the time when they are brightest.

As you see, the planets are, at times, as bright as or even brighter than any of the stars. No wonder their odd motions against the starry backgrounds were so easily detected by the early stargazers.

TABLE 3

Magnitudes of the Planets

PLANET	MAXIMUM MAGNITUDE
Mercury	−1.2
Venus	−4.3
Earth	—
Mars	−2.8
Jupiter	**−2.5**
Saturn	−0.4

When Mars is at its closest, it shines like a ruddy jewel in the sky with its magnitude of −2.8. It is then nearly four times as bright as Sirius, the brightest ordinary star. When Mars is farthest away, however, its magnitude is only +1.4 and there are over a dozen ordinary stars that are brighter than it is.

But now that we've seen how things are with Mars, how is it with Jupiter? By Kepler's model, Jupiter is 3.5 times as far from the sun as Mars is. When Earth, Mars, and Jupiter are all on the same side of the sun, Jupiter is over 8 times as far from us as Mars is.

When Jupiter is on the other side of the Earth, it is farther from us by the entire width of Earth's orbit. However, Jupiter is so far away from us that the entire width of Earth's orbit isn't much to add.

What it amounts to is that Mars is so close to us that adding Earth's orbit to the distance makes a big difference; Jupiter is so far that adding Earth's orbit makes much less difference. When Jupiter is on the other side of the sun from Earth, it is only 1.5 times as far away from us as when it is on the same side of the sun as Earth is.

This means that Jupiter's brightness changes much less than that of Mars does. When Jupiter is at its closest and brightest, its magnitude is —2.5. When it is at its most distant and dimmest, its brightness is —2.0. Jupiter is *always* considerably brighter than even the brightest stars and, except for a brief period when Mars is at its closest to us, it is always brighter than Mars.

(To be sure, when Mars and Jupiter are on the opposite side of the sun from us, the sun lies between us and them, and we can't see the planets. However, even though we can't see them, we can calculate what their brightness would be if we could see them.)

But this brings up an interesting point. If Jupiter is so much farther away from us than Mars is, why should Jupiter be the brighter of the two?

To answer that, we have to ask where the light of the planets comes from. Could it be that the planets burn and shine with their own light as the sun does? In that case, it might be that Jupiter simply burns with a brighter flame than Mars does.

But it seems doubtful that the planets are shining with their own light. The Earth is one of the planets and certainly has no light of its own. The side of the Earth that faces the sun is lit up and must shine as seen from space. The side of the Earth that is away from the sun is dark and does not shine.

That is the case with the moon, too. The ancient Greeks could see that it was the side of the moon that faced the sun that gleamed with light. The reason the moon showed different shapes ("phases") was that different parts of it (as seen from Earth) faced the sun.

The Earth and the moon, in other words, shine only by reflecting light from the sun.

Perhaps the other planets did the same. Perhaps they, like Earth and its moon, were spheres that caught the light of the sun and reflected it back into space. Perhaps it is just that the planets, being so far way, appear to us as very tiny spheres, too small to see as anything more than dots of light.

Suppose we imagine Mars and Jupiter as each being a sphere that reflects sunlight. Jupiter, however, is much farther from the sun than Mars is. Light traveling from the sun to Jupiter and then being reflected from Jupiter to us must travel a path that is 4.5 times as long, on the average, as the light that goes from the sun to Mars and then back to Earth.

Light fades out in proportion to the square of the distance. That means that if one beam of light travels 4.5 times as far as another (with both of equal brightness to start with), the amount of light falling on a particular area is 4.5 x 4.5 or 20.25 times more for the light traveling the shorter distance.

In other words, the light traveling the long distance from the sun to Jupiter to Earth should be dimmed so much that Mars should be twenty times as bright as Jupiter, when both are closest to us. Instead, Mars is just slightly brighter under those conditions.

Why should this be? Is Jupiter's sphere shinier than that of Mars so that Jupiter reflects more light? Is Jupiter's sphere larger than that of Mars so that Jupiter catches more light to reflect? This could not be decided from Kepler's model alone.

Something more was needed, and even when Kepler was beginning to work out his model, that "something more" was being brought into existence.

2

THE
LARGEST
PLANET

The Telescope to the Rescue

In 1609 the Italian scientist Galileo Galilei (usually known by his first name only) heard an interesting rumor. He was told that in the Netherlands someone had built a tube with lenses at each end, and that this tube made distant objects look closer when one looked through it. Galileo experimented with tubes and lenses and soon built such a device of his own. He called it a "telescope" from Greek words meaning "to see far."

Almost at once, Galileo turned his telescope on the heavens.

When he looked at the moon, it instantly appeared larger and brighter. Not only did his telescope magnify the moon in general, it also magnified every detail on its surface. Some objects which were too small to be seen with the unaided eye became visible through the telescope. Galileo, using his telescope, could see mountains and craters on the surface of the moon.

When he looked at the stars through the telescope, they appeared brighter. Indeed, he saw hundreds of stars that were too dim to be seen at all without a telescope. The dimmest stars that could be seen by eye alone were of magnitude 6. With the telescope, still dimmer stars could be seen, stars with magnitudes of 7, 8, and so on.

Even through the telescope, however, the stars remained

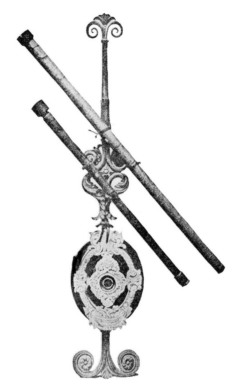

Nowadays, telescopes can be enormous devices that have to be moved by delicate machinery. The first telescopes, however, were simple spyglasses. These are two telescopes Galileo used, scarcely better than opera glasses, but with them he made marvelous discoveries—including the four large satellites of Jupiter.

nothing more than points of light. They were brighter points, but still points. If they were spheres, they were so small that even under the magnifying influence of the telescope they remained too small to be seen as anything more than dots.

But then Galileo looked at the planets. To his delight, they were not only brighter when seen by telescope, but they were enlarged into tiny spheres.

This made it possible to determine something new about the planets. In time to come, astronomers were able to measure, quite accurately, the width of a planetary globe as seen through the telescope. Knowing exactly by how much a telescope magnified objects, they could then calculate the width of the planet as it was without the telescopic magnification—the width as it appeared to the unaided eye.

Naturally, we measure the width of a heavenly body in degrees. The sun and the moon are each about 0.5° in width. The planets, however, are much smaller, in appearance, than the sun and moon are, and we would be working with inconveniently small decimals if we used degrees as our unit of measures.

As it happens, though, the Sumerians had divided each degree into sixty equal parts which we call "minutes of arc." We use (') as the symbol for a minute of arc, so we can say $1° = 60'$. Then, too, each minute of arc is divided into sixty equal parts which we call "seconds of arc" and symbolize (") so that $1' = 60''$.

Thus, instead of saying that the moon is about 0.5° in diameter we can say it is about 30' wide. If we want to be more accurate, we can say that the average width of the moon as seen from the surface of the Earth is 31' 5". We can put that width into seconds of arc, too, if we multiply 31 by 60 and then add the 5. It turns out that the average width of the moon is 1865".

The diameter of the sun as seen from the Earth is just a little greater than that of the moon. The sun is 31' 59" wide, or 1919" wide.

Well, then, how does the apparent diameter of the various planets, as determined by telescopic observation, compare with the apparent diameter of the moon? You will find the figures for each of the planets known to Galileo in Table 4. In each case, the maximum apparent diameter is given for the planet at its closest approach to Earth, and the minimum apparent diameter at its moment of farthest distance.

As you see in the table, Jupiter, when it is closest to Earth, has an apparent diameter of 50.0". It is only about 1/37 as wide, in appearance, as the moon. The globe of Jupiter, even at its

Galileo Galilei, commonly known as Galileo, was born in Pisa, Italy, in 1564. The first astronomer to use a telescope, he made astonishing discoveries which included the mountains on the moon, spots on the sun, and the phases of Venus. He also discovered the four largest satellites of Jupiter in 1610. He died in 1642.

maximum apparent width, is so small that to the eye it appears only like a dot of light.

But consider the case of Mars. When it is at its closest to Earth, its apparent width is 25.1″.

That's interesting. When both are as close as they can be to Earth, Jupiter is over 8 times as far away from us as Mars is. Things look smaller as their distance increases, and if Jupiter and Mars were roughly the same size, Jupiter would look considerably the smaller of the two because of its greater

TABLE 4

Apparent Width of the Planets

PLANET	DIAMETER (IN SECONDS OF ARC)	
	MAXIMUM	MINIMUM
Mercury	12.7	4.7
Venus	64.5	9.9
Earth	—	—
Mars	25.1	3.5
Jupiter	**50.0**	**30.8**
Saturn	20.6	14.9

distance. This, however, is not so. Despite Jupiter's considerably greater distance, its apparent diameter is twice that of Mars.

To account for Jupiter's greater apparent size even though it is so much farther away than Mars is, we must accept the fact that Jupiter has a diameter much greater than that of Mars. Indeed, judging by what could be seen through the telescope, Jupiter was the largest of the planets to be seen in the sky.

At least it was larger than any of the planets that were known in the time of Galileo. As it happens, though, three additional planets have been discovered since Galileo's time. What of those?

In 1781 an English astronomer, William Herschel, discovered a seventh planet, and this was named Uranus after the god of the sky in the Greek myths. In 1846 a French astronomer, Urbain Leverrier, discovered an eighth planet, which was named Neptune after the god of the sea. Finally, in 1930 an American astronomer, Clyde Tombaugh, detected a ninth

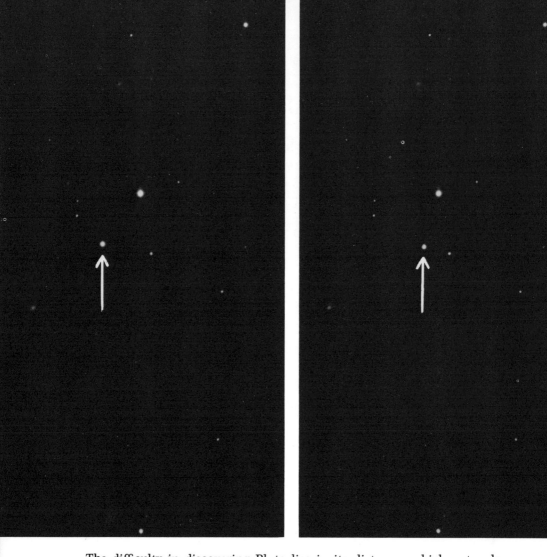

The difficulty in discovering Pluto lies in its distance, which not only makes it dim so that it is lost among the stars, but slow-moving as well so that it can't be easily distinguished from them. Once spotted, though, the movement is plain if a large telescope is used. This photograph shows the motion of Pluto in 24 hours, as seen through the 200-inch telescope at Mount Palomar.

planet which was named Pluto after the god of the underworld.

To bring us up to date on these new planets, Table 5 contains data in connection with them, data of a kind already given for the other planets in earlier tables.

If we consider the eccentricity and inclination of the orbits of Uranus and Neptune, we see that these seem typical planets. Pluto's orbit, on the other hand, is most unusual. Its orbit is more eccentric and has a higher inclination than that of any other planet. For that reason alone it deserves a book of its own, but in this book, which concentrates on Jupiter, Pluto can only be mentioned in passing.

The magnitude of Uranus, at its brightest, is 5.7. Unlike the planets known in ancient times, Uranus is quite dim. It is just barely bright enough to see. No wonder the ancients didn't

TABLE 5

Planets Discovered in Modern Times

PLANET	ECCEN- TRICITY	INCLINA- TION (IN DEGREES)	MAGNITUDE (MAXIMUM)	APPARENT WIDTH IN SECONDS (MAXIMUM)
Uranus	0.047	0.8	5.7	4.2
Neptune	0.009	1.8	7.6	2.4
Pluto	0.249	17.1	14.5	under 1

notice that a very dim star was moving slowly across the pattern of the stars and therefore had to be considered a planet. It took the magnifying and brightening effect of the telescope to make Uranus clearly visible and to make its motion obvious.

As for Neptune and Pluto, they are even dimmer than Uranus and are absolutely invisible to the unaided eye. Without a telescope, they could never be seen.

The reason for the unusual dimness of these three planets is a simple one. They are all more distant from the sun (and from us) than any of the longer-known planets.

From the rate at which they move against the starry background of the sky, it could be shown that Uranus is twice as far from the sun as Saturn is. Neptune is three times as far from the sun as Saturn is, and Pluto is four times as far.

From the apparent size of the new planets as seen in the telescope and from their position in the solar system, it is easily seen that not one of them can rival Jupiter in size.

Suppose we let the diameter of Jupiter be represented by 50. The diameters of the other planets, including the three not known in the time of Kepler and Galileo, can then be represented by other numbers in proportion. Such relative diameters are given in Table 6.

Parallax to the Rescue

Notice that the list of planets in Table 6 does not include Earth.

If all we have is Kepler's model, then we can work out the relative sizes of the other planets, but not their absolute sizes. We can say that Jupiter has three times the diameter of

TABLE 6

Relative Diameters of the Planets

PLANET	RELATIVE DIAMETER (JUPITER = 50)
Jupiter	**50**
Saturn	42
Uranus	17
Neptune	16
Venus	4.3
Mars	2.4
Pluto	2.3
Mercury	1.7

Neptune, but we can't say what the diameter is, in miles, for either one. Jupiter might have a diameter of 900 miles and Neptune one of 300; or Jupiter might have a diameter of 9,000,000 miles and Neptune one of 3,000,000. In either case Jupiter has three times the diameter of Neptune, but which one is correct, if either is, we have no way of telling.

What can we tell from the fact that Jupiter has a diameter of 50″ as seen from Earth? If Jupiter were quite close to us, it wouldn't have to be very large to appear as a globe 50″ in diameter.

For instance, an object the size of the Earth (which is 8000 miles in diameter) would be seen as a globe 50″ in diameter if it were 32,000,000 miles away from us. To be seen as 50″ in diameter from a greater distance, the object would have to be larger than the Earth; to be seen as 50″ in diameter from a smaller distance, the object would have to be smaller than the Earth.

For all Galileo knew, Jupiter might be closer than 32,000,000 miles from Earth at its closest and might therefore be smaller than the Earth. The other planets would, of course, be smaller still, so that Earth might prove the giant of the solar system.

That could be—but was it? If the scale of the solar system were large enough and if all the planets were *very* far away, even Mercury might be larger than the Earth in actual fact, and we might be inhabiting the pygmy of the solar system.

Any consideration of the properties of the planets depends, therefore, entirely on the scale of the solar system—on knowing how far away the planets are.

If the distance of even a single planet in the solar system could be determined, then the distances of all the rest could also be determined from Kepler's model. But how do we go about determining the distance of that all-important one planet to begin with?

The only known way, in Galileo's time, of measuring the

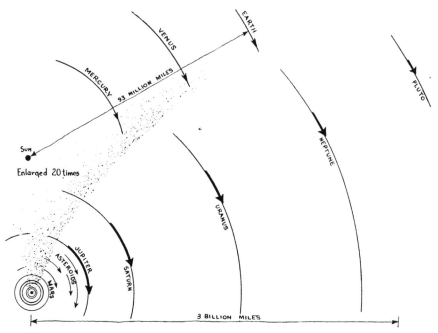

The Solar System

Scale of Planet Sizes

Diagrams are limited, of course. Here Jupiter doesn't look much larger than Saturn (especially when the rings are included for the latter), since it isn't—in diameter. However, Jupiter is three times as large as Saturn in mass. As for the planetary orbits, they are not perfect circles, so although the *average* distance of Pluto is far beyond Neptune as shown, there are times when Pluto is actually slightly closer to the sun than Neptune is.

distances of objects in the sky was by noting their shift in position when viewed from different places.

To see this, hold your finger up in front of your eyes. Now look at it with one eye at a time. First close your right eye and look at it with your left eye, noticing its position against some object in the background. If you are indoors, note its position against the wall. It may be near the edge of a picture.

Now, while you hold the finger motionless, close your left eye and look at it with your right eye. The finger has changed position!

This shift in position is called "parallax," and the size of the parallax depends on distance. If you hold your finger quite close to your eyes, the shift in position will be great. If you move your finger farther away from your eye, the shift gets smaller.

From the amount of the parallax, you can calculate the distance of your finger from your eye by making use of trigonometry, a branch of mathematics.

Just as your finger shows a parallax against the more distant wall, so the moon shows a parallax against the more distant stars. The moon is so much more distant than your finger, however, that the parallax is a tiny one indeed. If you tried to look at the moon first with your right eye and then with your left eye, you would not see any change in the position of the stars that may happen to be near the moon.

In this case, though, we are viewing the moon from two positions (each of your two eyes) which are separated by only a few inches. Suppose that you look at the moon from an observatory near Los Angeles while another astronomer looks at it, at the same time, from an observatory in New York. The difference in positions of observation would then be nearly 3,000 miles and the parallax would become large enough to measure.

In this way, the actual distance of the moon can be determined. It turns out that the average distance from the center of the moon to the center of the Earth is just under 239,000 miles. The moon, however, is not part of Kepler's model so you can't work out the planetary distances from the distance of the moon.

The planets in the sky are all much more distant than the moon. They have parallaxes that are so small that even a change in the observer's position of thousands of miles doesn't make them large enough to measure with the unaided eye. That is why the distance of the sun from the Earth, or the distance of any of the planets, couldn't be worked out even in Galileo's time.

But what if one used a telescope? The magnifying power of a telescope could make a tiny shift in position large enough to measure. With a telescope good enough, tiny parallaxes could be measured.

In 1671 the parallax of a planet was measured for the first time. In Paris, France, an Italian-French astronomer, Giovanni Domenico Cassini, made careful measurements of the position of Mars compared to that of nearby stars at certain definite times. On the northern shore of South America another French astronomer, Jean Richer, made similar measurements.

Cassini compared his measurements with those of Richer when the latter returned to Paris, and from these worked out the parallax. From this he determined the distance of Mars and, therefore, all the other planetary distances in the solar system.

For the first time men learned the scale of the solar system.

Since Cassini's time, measurements of parallaxes have been made more accurately, and other systems of determining distances have been used, but Cassini gets the credit for the first determination.

In Table 7 the average distance from the sun of each of the nine planets is given in three ways.

First, it is given in miles, which is the most common unit of measurement for long distances here in the United States.

Second, it is given in kilometers, which is the unit of measurement for long distances used in almost the entire world outside the United States. The kilometer is the smaller unit of the two, being equal to about ⅝ of a mile. It is a unit of measure in the metric system, and all the tables in this book will give metric measurements as well as those common in the United States.

Third, it is given in astronomical units or "A.U.," where an A.U. is equal to the average distance of the Earth from the sun. By using this unit, you can tell at once how the distance of a particular planet from the sun compares with that of the Earth's distance from the sun.

Notice that in Table 7 it is the *average* distance of the planets from the sun that is given. If the planets traveled about the sun in a perfect circle, the distance from the sun would be the same at all times for each planet. The planets travel in ellipses, however, with the sun at one focus, and this means the sun is at varying distances. The amount of change is not great in the case of most planets since the eccentricity is small (see Table 1 on page 23), but it is there.

The Earth, for instance, is at an average distance from the sun of 93,000,000 miles. When it is at perihelion, however, at the point in its orbit closest to the sun, it is only 91,400,000 miles away. At the opposite point in its orbit, when it is at

TABLE 7

Average Distances of the Planets From the Sun

| PLANET | AVERAGE DISTANCE | | |
	IN MILLIONS OF MILES	IN MILLIONS OF KILOMETERS	IN ASTRONOMICAL UNITS (A.U.)
Mercury	36	58	0.39
Venus	67	108	0.72
Earth	93	150	1.00
Mars	141	228	1.52
Jupiter	**483**	**779**	**5.20**
Saturn	886	1428	9.54
Uranus	1782	2872	19.2
Neptune	2792	4501	30.0
Pluto	3671	5906	39.5

aphelion and farthest from the sun, it is 94,600,000 miles away. This is a difference of 3,200,000 miles, but compared with the total size of the Earth's orbit it isn't very much.

Naturally, the size of the sun changes as the Earth moves in its orbit. The sun appears larger when the Earth comes closer, and appears smaller when the Earth moves off farther. The average width of the sun is 31′ 59″. When Earth is closest to the sun, however, the sun has a diameter of 32′ 30″, and when Earth is farthest from it, it has a diameter of 31′ 28″. The difference isn't noticeable except by careful measurement. The sun looks the same size at all times.

In Table 8 the maximum and minimum distances of each planet from the sun are given in astronomical units. Only Mercury and Pluto, which have the greatest orbital eccentricities of any of the planets, show really large differences in distances.

For Jupiter, a moderate eccentricity means that it is sometimes as much as 507,000,000 miles from the sun, and sometimes as close as 460,000,000 miles from it. The closest approach of Jupiter to Earth is when Jupiter is at perihelion and the Earth is on the same side of the sun as Jupiter is. Jupiter can then be as little as 365,000,000 miles from Earth. When Jupiter is at aphelion and Earth is on the other side of the sun, Jupiter can be as far from us as 600,000,000 miles.

Planetary Motion

Once the scale of the solar system was determined, a number of interesting properties of the planets could be calculated.

Suppose we begin, for instance, with the fact that the Earth goes around the sun once in 365¼ days. This period of time is called a year.

Allowing for the motion of the Earth, astronomers can use the motion of the planets in the sky to work out the length of time it takes each of them to make one complete circle about the sun. The time it takes for a planet to do this is its "period of revolution."

TABLE 8

Maximum and Minimum
Distances of the Planets From the Sun

| PLANET | DISTANCE FROM THE SUN (IN A.U.) | |
	MAXIMUM	MINIMUM
Mercury	0.47	0.31
Venus	0.72	0.72
Earth	1.02	0.98
Mars	1.66	1.39
Jupiter	**5.45**	**4.95**
Saturn	10.1	9.0
Uranus	20.0	18.3
Neptune	30.2	29.9
Pluto	48.3	29.8

It turns out, as you can see in Table 9, that the farther away a planet is from the sun, the longer it takes for it to make the complete circuit. This is certainly not surprising, since the farther a planet is from the sun, the larger the circle it must make in going around it.

Notice, by the way, that there is a big jump in period of revolution between Mars and Jupiter. The first four planets all have periods of revolution of less than two years, so that it is often convenient to give those periods in days, as is shown in Table 9. The last five planets, on the other hand, all have periods of revolution of over eleven years.

This is one of the sharp differences between the two groups of planets which causes astronomers to refer to Mercury, Venus, Earth and Mars, in a group, as the "inner planets." Jupiter and the planets beyond are the "outer planets." (There are other differences between the two groups as well, and we will come to those.)

The fact that a more distant planet has a longer period of revolution than a closer one does is not a matter of distance only, however. The more distant planet not only has a longer distance to travel over in making its circle around the sun, it moves more slowly as well.

We can show this by making use of the scale of the solar system to calculate the actual speed at which planets are moving around the sun.

For instance, Earth is at a distance of 93,000,000 miles from

TABLE 9

Period of Revolution of the Planets

PLANET	PERIOD OF REVOLUTION	
	YEARS	DAYS
Mercury	0.24	88
Venus	0.62	225
Earth	1.00	365
Mars	1.88	687
Jupiter	**11.86**	
Saturn	29.46	
Uranus	84.02	
Neptune	164.8	
Pluto	248.4	

the sun so the length of its orbit is 585,000,000 miles. (In order to get the length of the orbit, you must multiply the distance from the sun by 6.28. This is really correct only if the orbit is a circle, but the elliptical orbits of the planets are so close to a circle that the multiplication gives us fairly accurate results.)

The Earth covers that 585,000,000-mile journey in exactly one year. In order to do so (considering that there are 31.5 million seconds in a year), it has to travel at an average speed of 18.5 miles per second as it circles the sun. (The Earth moves a little more quickly at the perihelion end of its orbit when it is closer than average to the sun, and a little more slowly at the aphelion end, but the differences aren't much.)

Jupiter is 5.2 times as far from the sun as the Earth is. That means its orbital length is 5.2 times as long as Earth's is. Jupiter's orbit is about 3,000,000,000 miles long. If it traveled at Earth's speed it would complete the trip about the sun in 5.2 years.

But it doesn't; it takes longer. Jupiter takes nearly 12 years to do the job. Its average orbital speed is less than half that of Earth, therefore. In Table 10, the average orbital speed of all nine planets is given. From the table you can see that the farther a planet is from the sun, the more slowly it moves in its orbit.

The reason for this is that a planet moves about the sun in response to the grip of the sun's force of attraction. This force of attraction is produced by the sun's gravitational field, and this field gets steadily weaker with distance. A planet that is far away from the sun is influenced by a weak gravitational field and therefore moves more slowly than a planet nearer to the sun which is therefore exposed to a stronger gravitational field.

In 1687, the English scientist Isaac Newton worked out the exact manner in which the gravitational field grew weaker with distance. He showed that the weakness changed according to the square of the distance. (You get the square of a number by multiplying it by itself, so that the square of 2 is 4, since $2 \times 2 = 4$. In the same way, the square of 3 is 9, the square of 4 is 16, and so forth.)

TABLE 10

Average Orbital Speed of the Planets

| | AVERAGE ORBITAL SPEED | |
| | MILES | KILOMETERS |
PLANET	PER SECOND	PER SECOND
Mercury	29.8	47.9
Venus	21.8	35.1
Earth	18.5	29.8
Mars	15.0	24.1
Jupiter	**8.1**	**13.1**
Saturn	6.0	9.6
Uranus	4.2	6.8
Neptune	3.4	5.4
Pluto	2.9	4.7

Suppose, then, that the sun's gravitational field has a particular strength at a distance of 100 million miles from the sun. At 200 million miles the distance has been multiplied by 2, and the strength of the gravitational field has dropped to ¼. At 300 million miles the distance has been multiplied by 3 and the strength of the gravitational field has dropped to ⅑. At 400 million miles, the distance has been multiplied by 4 and the strength of the gravitational field has dropped to ¹⁄₁₆.

Newton showed that if gravitational fields behaved in this manner, the planets had to move in ellipses with the sun at one focus; that a planet had to speed up as it moved closer to the sun, and slow down as it moved farther from the sun; that the orbital speeds of planets at various distances from the sun had to be exactly what astronomers had found them to be.

It was this slow motion of the outer planets, by the way, that

Isaac Newton was born in Lincolnshire, England, in 1642. He was probably the greatest scientist who ever lived, and his crowning achievement was to work out the law of universal gravitation in 1687. This made it possible to understand how the planets went around the sun, how the satellites went around the planets, down to very fine details. He died in 1727.

helped make it difficult to locate the planets beyond Saturn. Table 11 gives the average angular distance through which each of the outer planets moves across the sky in one day, and how long it would take each one to move across a distance equal to the width of the moon.

Jupiter moves against the stars through a distance equal to

the width of the moon in a little over 6 days, while Saturn does it in a little over 15 days. As the early stargazers watched night after night, this motion became quite obvious—especially for

TABLE 11

Apparent Motion of the Outer Planets

PLANET	APPARENT MOTION IN 24 HOURS (IN SECONDS OF ARC)	TIME TO MOVE THE WIDTH OF THE MOON (IN DAYS)
Jupiter	**300**	**6.2**
Saturn	120	15.5
Uranus	42	44
Neptune	22	85
Pluto	14	130

bright objects like Jupiter and Saturn that attract the eye.

Uranus is a very dim object, however. What's more, it takes an average of six weeks for it to move the width of the full moon. Even if someone noted Uranus, there was no reason to watch such a dim and uninteresting object for night after night. And if someone did, it would seem to stay in much the same place for several nights in a row and the stargazer would lose interest. In the end, Uranus was discovered to be a planet by sheer accident. Herschel happened to look at it through his telescope, and he noticed it appeared to be a globe instead of a mere dot of light.

Neptune and Pluto were even more difficult to find because they are not only considerably dimmer than Uranus, but move even more slowly. It takes Pluto over three months to move a distance equal to the moon.

Planetary Size

Jupiter is in the middle of the planetary group, being fifth out of nine. There are four planets closer to the sun than Jupiter is, and four planets farther. This intermediate position in distance makes Jupiter intermediate in everything that depends on distance. It is intermediate in the length of its period of revolution, intermediate in its average orbital speed, intermediate in the speed with which it moves across Earth's sky.

It is not intermediate in everything, however. Remember, it is larger than any of the other planets in the sky, and only the relationship of its size to Earth's size remained uncertain in Galileo's time. This last relationship could be determined once the scale of the solar system was worked out.

It turned out that when Jupiter was at its maximum apparent size of 50″, it was 360 million miles from Earth. In order to seem that size when seen at that distance, it was necessary for Jupiter to be larger than anyone had expected. It was, in fact, much larger than Earth.

In Table 12 the diameters of the various planets are given as measured across the equator.

As you see, Jupiter is not in an intermediate position in this respect. It is at an extreme, for it is the largest planet in the solar system. Its diameter is 11.2 times that of the Earth. In other words, if eleven spheres the size of the Earth were placed side by side they would not quite stretch across Jupiter's great globe.

We can go further than that, too, for the figures for the diameter of the various planets don't really tell us by how much Jupiter is the largest planet.

What is important to us about the Earth, for instance, isn't so much its diameter as its surface area. After all, we live on the surface.

The surface area of any sphere is proportional to the square

TABLE 12

Diameter of the Planets

PLANET	DIAMETER AT THE EQUATOR		
	IN MILES	IN KILOMETERS	EARTH = I
Mercury	3,100	5,000	0.39
Venus	7,570	12,200	0.95
Earth	7,927	12,757	1.00
Mars	4,200	6,750	0.53
Jupiter	**88,700**	**142,900**	**11.2**
Saturn	75,100	120,900	9.5
Uranus	29,000	46,500	3.6
Neptune	28,000	45,000	3.5
Pluto	4,000	6,500	0.5

TABLE 13

Surface Area of the Planets

PLANET	SURFACE AREA (EARTH = I)
Mercury	0.15
Venus	0.90
Earth	1.00
Mars	0.28
Jupiter	**125**
Saturn	90
Uranus	13
Neptune	12
Pluto	0.25

of its diameter. This means that if Sphere A has 2 times the diameter of Sphere B, it has 2 × 2 or 4 times the surface area of Sphere B. If Sphere A has 10 times the diameter of Sphere B, it has 10 × 10 or 100 times the surface area.

Suppose, then, we consider the surface area of the Earth (which is equal to 200,000,000 square miles) to be equal to 1. What will be the surface area of the other planets in relation to that? The answer is given in Table 13.

(Of course, when we talk about surface area, we are pretending that all the planets are like Earth with a nice solid surface. There will be more to say about that later in the book, but for now, let's pretend this is so.)

As you see from Table 13, Jupiter has a surface area that is as large as that of all the other eight planets put together. In fact, if you could imagine the surface of the Earth spread out and pasted onto Jupiter, it would cover about half as much of Jupiter as the United States does of the Earth.

Let's see how else we can look at Jupiter's size. We might wonder how much space it occupies. A sphere has a certain diameter and a certain surface area, but it also takes up a certain amount of space; that is, it has a certain volume.

The volume of a sphere varies as the cube of its diameter. In other words, if Sphere A has 2 times the diameter of Sphere B, then Sphere A has 2 × 2 × 2 or 8 times the volume of Sphere B. From the diameter of each planet we can calculate its volume and see how that compares to the volume of the Earth. This is done in Table 14.

From this viewpoint, Jupiter seems larger than ever. The volume of Jupiter is nearly twice that of the next largest planet, Saturn. The volume of Jupiter is more than one and a half times as large as the volume of all the other planets put together. If you could imagine Jupiter as completely hollow, you could pack 1400 spheres the size of the Earth into it.

In terms of the size of the planets, as you can see from Tables 12, 13, and 14, it is convenient to divide the planets into two groups. The four inner planets, plus Pluto, are each less than 8,000 miles in diameter. Of these five planets, Earth is largest, and they are therefore grouped together as the "Terrestrial planets," from the Latin word for Earth.

Jupiter, Saturn, Uranus, and Neptune are all over 25,000 miles in diameter and, of these, Jupiter is the largest. These four planets are grouped together, therefore, as the "Jovian

TABLE 14

Volume of the Planets

PLANET	VOLUME (EARTH = 1)
Mercury	0.06
Venus	0.86
Earth	1.00
Mars	0.15
Jupiter	**1400**
Saturn	860
Uranus	47
Neptune	43
Pluto	0.12

planets," from Jove, which is an alternate name for Jupiter.

The division of the planets into a Terrestrial group and a Jovian group is a useful one because, as we shall see, there are many other differences besides the one of mere size.

Yet although Jupiter is far and away the largest planet in the solar system, it isn't the largest *object* in the solar system. The sun is 93,000,000 miles from Earth, which is only a quarter the distance of Jupiter at its closest approach to us, so it would be no great surprise if the sun appeared larger than Jupiter.

Yet it appears so *much* larger. Its apparent diameter is 38 times that of Jupiter.

Taking into account the apparent size of the sun and its distance from us, it turns out that the sun is a sphere with a diameter of 865,000 miles. This is 109 times the diameter of the Earth and is nearly 10 times the diameter of Jupiter. The width of the sun compared to Jupiter is very nearly that of Jupiter compared to the Earth. The sizes of Jupiter and the sun are compared in Table 15.

The great size of Jupiter compared with that of other planets answers the question of its brightness, at least in part.

It is farther from the sun than Mars is and gets dimmer light, to begin with. It is also farther from the Earth than Mars is, so that light reflected from Jupiter has farther to go and gets dimmer than light reflected from Mars. Nevertheless, Jupiter appears brighter to us than Mars does because there is so much of the giant planet. Jupiter's broad globe catches far more light than tiny Mars does.

We can make another point, too, in this connection. Not every object reflects the same fraction of the light it receives. A dull, dark rock absorbs most of the light that strikes it and then gives it off in the form of invisible infrared rays. It reflects very little visible light; that is why it looks dull and dark. A shiny object reflects a great deal of the light that falls on it, in the form of visible light; that's why it looks shiny.

What about the planets? Once their distances from the sun is known, and their size, it is possible to calculate how bright they would seem if they reflected *all* the sunlight that fell on them. From their actual brightness we can tell what fraction of the light they reflect.

The moon, for instance, has no atmosphere. Light that falls on the moon strikes the dull rocks that make up its surface. The moon only reflects 0.06 of the light that strikes it. This figure is called the moon's "albedo" (from a Latin word meaning "whiteness").

TABLE 15

Jupiter and the Sun Compared

	JUPITER	SUN
Diameter	1	9.8
Surface area	1	95
Volume	1	925

If the moon had an atmosphere, the air would reflect some of the light better than the bare rocks of the surface would. Any clouds that formed in its air would reflect light particularly well. From the albedos of the various planets, given in Table 16, something can be told about the nature of the surface.

Since Mercury reflects light no better than the moon does, it probably lacks an atmosphere, too. Mars does have an atmosphere. That can be seen by telescope, and sometimes it even has a haze in it. Mars's albedo is two and a half times that of the moon.

The earth has a thicker atmosphere than Mars does and has more clouds. Satellites have measured the amount of sunlight the Earth reflects, and its albedo turns out to be 0.3, fives times that of the moon and twice that of Mars.

Those planets that have higher albedos than Earth does must have atmospheres with thicker clouds. Venus, for instance, which has the highest albedo of any planet, is covered with an eternal layer of white clouds. All we can see of Venus through the telescope is a featureless expanse of white.

In the same way there are unbroken clouds hovering eternally over the Jovian planets. We never see the actual solid surface of Jupiter, for instance, any more than we see that of Venus. The Jovian planets have albedos somewhat less than that of Venus, however. This must be because the clouds on

TABLE 16

Albedos of the Planets

PLANET	ALBEDO
Mercury	0.07
Venus	0.59
Earth	0.30
Mars	0.15
Jupiter	**0.44**
Saturn	0.42
Uranus	0.45
Neptune	0.52
Pluto	?

NOTE: *Pluto is so small and so far away that it is far less well known than the other planets. In some tables I can give no figures for it (as in this one), and even when I do give figures, they are rarely as reliable as are the figures for the other planets.*

the Jovian planets are not quite the same in nature as those on Venus, and don't reflect light quite as well. (We'll get back to this later in the book.)

The fact that Jupiter's albedo is three times that of Mars also contributes to the great brightness of Jupiter.

3

THE
SHAPE
OF
JUPITER

Spinning Spheres

Let's go back, now, to the early studies of Jupiter by telescope.

In 1691 Cassini, who first worked out the scale of the solar system, was studying Jupiter. He was using a better telescope than Galileo had had, and it was quite plain to him that Jupiter was not a perfect circle of light. That meant its planetary shape could not be that of a perfect sphere.

Actually Jupiter's outline is that of an ellipse, so that its shape must be that of a sphere that has been flattened from two opposite sides—rather like a tangerine. This shape is called, by mathematicians, an "oblate spheroid."

This was rather astonishing when it was noticed. The only bodies in the heaven that were more than a point of light to the unaided eye were the sun and moon, and both seemed perfect circles of light. That meant they were perfect spheres. And it had long been thought that the Earth was a perfect sphere, too.

Why should Jupiter be different?

The explanation can be worked out from Isaac Newton's law of gravitation. It is the gravitational pull of a body on its own substance that makes it a sphere. A small body—say, a boulder a mile across, floating in space—has such feeble gravity that it can have any shape, however irregular. A large body, one

Giovanni Domenico Cassini was born near Nice in 1625. This was then Italian territory, but he did his work in astronomy in Paris. He studied the period of Jupiter's revolutions and the period of the revolution of its satellites. He was the first to notice the shadows cast by Jupiter's satellites on its surface, the first to notice that Jupiter was oblate, and, by working out the scale of the solar system, was also the first to determine Jupiter's distance. He died in 1712.

that is over a hundred miles in diameter, perhaps has a strong enough gravity to pull all parts of itself as close to its center as possible. When all parts of a body are drawn as close to the center as possible, that body has the shape of a sphere.

Of course there can remain some unevennesses. There are mountains and valleys on the surface of the Earth, for instance. These may look very uneven indeed, but even the highest mountain has a height that is only $\frac{1}{3000}$ the diameter of the Earth. If you made a model of the Earth the size of a billiard ball, and if the mountains and all the other unevennesses were placed on it exactly to scale, all the unevennesses would be so small that the Earth would end up smoother than an ordinary billiard ball.

Is there anything that can fight gravity and keep a large body from being perfectly spherical?

Yes!

Suppose the body rotates. If it rotates, it does so about some central line or axis. This axis would reach the surface of the body at the poles. Thus, the Earth rotates about an axis which reaches the surface at Earth's north pole and south pole. A line around the Earth exactly halfway between the two poles is the equator.

As the Earth rotates about its axis, different parts of its surface move in different ways. A point just at the poles won't move as the Earth rotates. A point a small distance from the poles will move in a small circle as the Earth rotates. A point farther from the poles will make a larger circle, and a point at the equator will make the largest circle of all.

Since the Earth turns all in one piece, a point on Earth's surface makes one complete circle in the same time—twenty-four hours—whether that circle is a tiny one near the pole, or a huge one at the equator.

If you live in Leningrad, U.S.S.R., for instance, which is 2,000 miles from the north pole, you will make a circle around the Earth which is 12,500 miles long. You will cover this distance in just twenty-four hours, so that you will be moving at 500 miles an hour.

If you live in Philadelphia, U.S.A., which is 3,500 miles from the north pole, you will cover a distance of 19,000 miles in twenty-four hours. You will be moving at a speed of 800 mlies an hour.

Finally, if you are living in Quito, Ecuador, which is on the equator, 6,200 miles from the north pole, you will cover a distance of 25,000 miles in twenty-four hours and will be moving at a speed of just over 1,000 miles an hour.

When something is moving in a circle about some center, there is a tendency for it to be forced away from the center. (If you have a little ball with a thin rubber string attached, and whirl that ball around your head, the string will stretch as the ball is forced away from the center.) This is called a "centrifugal effect." The faster you rotate an object, the greater the centrifugal effect, and the greater the distance it moves away from the center.

As the Earth rotates, its substance is forced away from the central axis. Since the surface moves faster and faster the farther it is from the pole and moves fastest at the equator, the centrifugal effect is greater and greater the farther the place is from the pole, and is greatest at the equator. The substance of the Earth is lifted above where it would ordinarily be because of the centrifugal effect. Near the pole the lifting is very slight, but it gets larger and larger as one moves farther from the pole and is greatest at the equator.

For that reason, there is a bulge of matter around the Earth, stretching from pole to pole, and thickest at the equator.

How high is this equatorial bulge? If the surface of the Earth is lifted only an inch by the centrifugal effect (and less than that elsewhere), it would never be noticed.

Newton, using his law of gravitation, however, calculated that the equatorial bulge would be high enough to measure if a very careful survey of the Earth's surface were made. Such measurements were made in the 1730's, a decade after Newton had died, and he turned out to be correct.

The Earth's bulge due to the centrifugal effect is miles high in places. At the equator, the surface is 13.5 miles farther from the center than it would be if the Earth were not rotating.

This means that the Earth is an oblate spheroid. If you drew a circle completely around the Earth, from north pole to south pole and back to north pole, the result would be an ellipse, not a circle. You can say that the Earth bulges at the equator or that it is flattened at the poles. Either way of saying it is correct; in either case, the Earth is not a perfect sphere.

As a result, the diameter of the Earth from one point on the equator to an opposite point on the equator is slightly greater than the diameter from the north pole to the south pole. The equatorial diameter is 7,927 miles, while the polar diameter is only 7,900 miles. (It is for this reason that Table 12 gives the diameter of the planets at the equator. That is always the maximum diameter.)

The difference in the two diameters (27 miles) is only $\frac{1}{294}$ of the equatorial diameter. This fraction (or, in decimals, 0.0034) is called the "oblateness" of the Earth. Actually, the oblateness of the Earth isn't much. It certainly can't be seen at a casual glance. If you were to see the Earth from the moon, it would seem like a perfect circle to you. You would not notice that it had the slightest bulge around its middle or see the slightest flattening at its poles. Still, the oblateness is there and it can be measured.

But then why does the moon lack any detectable oblateness at all? Is it that it doesn't rotate?

No, it *does* rotate, but not quickly. The moon rotates once with respect to the stars in the same time that it revolves once about the Earth. This means that its period of rotation is 656 hours. The moon's circumference is 6,800 miles. A point on its equator would take 656 hours to move 6,800 miles and would be traveling at 10.4 miles an hour, compared to a motion of over 1,000 miles an hour at earth's equator. The centrifugal effect is therefore feeble indeed on the moon, and it is not surprising that it has no detectable equatorial bulge.

The sun rotates also. Its period of rotation is 601 hours, almost as long as that of the moon. On the other hand, the sun is an enormous body and its circumference is about 2,700,000 miles. At its equator, its surface is moving at a speed of 4,500 miles an hour, over four times the speed at which the Earth's equatorial surface is moving.

The sun's gravitational pull, however, is much greater than the Earth's. The centrifugal effect on the sun isn't enough to produce a detectable equatorial bulge against the sun's giant gravity. So the sun is a perfect sphere.

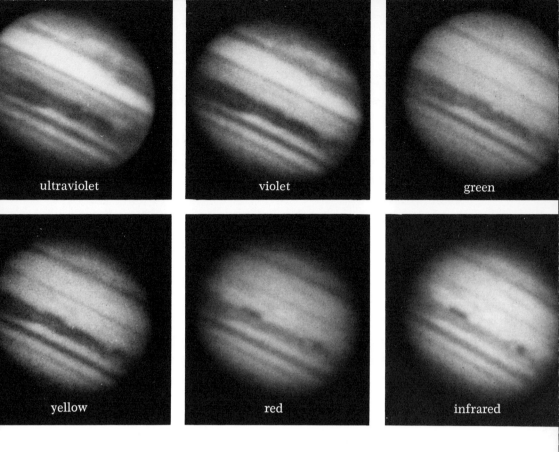

ultraviolet

violet

green

yellow

red

infrared

Jupiter can be photographed in different wavelengths of light, and in all, the bands and zones of Jupiter's surface can be seen. Note that Jupiter's outline is elliptical, not circular.

Bulge at the Equator

But what about the planets? Table 17 gives the oblateness of each, obtained by careful measurements of the telescopic image.

Mercury and Venus have practically no equatorial bulge at all. As nearly as we can tell, they are perfect spheres. That alone tells us that they probably rotate at a very slow speed. On the other hand, the Jovian planets are very oblate (another difference between them and the Terrestrial planets).

Jupiter's oblateness is 0.062. That means the difference between the equatorial diameter and the polar diameter is 0.062 of the equatorial diameter. The equatorial diameter is 88,700, and 0.062 of that is 5,400 miles. The polar diameter is therefore 5,400 miles less than the equatorial diameter and is only 83,300 miles.

Saturn is even more oblate than Jupiter. Its equatorial diameter is 75,100 miles, and 0.096 of that is 7,200 miles, so its polar diameter is 67,900 miles.

From the oblateness of the Jovian planets we can suppose that they are rotating, and quite quickly.

Cassini, who had reported Jupiter's oblateness in 1691, had also determined its period of rotation in 1665, twenty-six years earlier. He had done this by carefully noting certain spots he could detect on Jupiter's globe. He noticed that the spots kept drifting across the face of the globe till they disappeared around the other side. Eventually they would reappear on the side of the globe opposite from that where they had disappeared. It seemed quite obvious that they were being carried around Jupiter's globe by its rotation.

Watching night after night, he finally decided that Jupiter

TABLE 17

Oblateness of the Planets

PLANET	OBLATENESS
Mercury	0.000
Venus	0.000
Earth	0.0033
Mars	0.0052
Jupiter	**0.062**
Saturn	0.096
Uranus	0.06
Neptune	0.02
Pluto	?

rotated about its axis in just under ten hours. (It was in similar fashion that the period of rotation of the sun was determined. In that case, it was the passage of sunspots across and behind the surface of the sun that was used.)

Since Cassini's time, more sophisticated means of determining periods of rotation have been worked out. Table 18 gives the period of rotation of the various planets in hours and, in the case of the three planets with the longest periods, days as well. (As an indication, incidentally, that some apparently simple facts about long-known planets weren't learned till recently, the periods of rotation of Mercury and Venus were worked out in the 1950s and 1960s.)

Jupiter, as you can see from Table 18, turns on its axis in a shorter period than is true for any other planet. This is rather astonishing considering its size. Jupiter is eleven times as wide as Earth and therefore has a circumference eleven times as great, yet it rotates about its axis in less than half the time Earth takes. Clearly, Jupiter's equatorial surface must move much more rapidly than Earth's.

Table 19 gives the speed of rotation at the equator for each of the planets, and you can see how extreme Jupiter's position is in this respect. Saturn's equator moves almost as rapidly, but all the other planets lag far behind.

With the crawl of the equatorial surface of Mercury and Venus, we need not be surprised that they remain without detectable equatorial bulges. Though we can't study faraway Pluto's globe in order to measure its oblateness, we can be quite sure that it, too, has no equatorial bulge to speak of.

At the speed of Jupiter's equatorial surface, however, the centrifugal effect is enormous. Even considering the fact that Jupiter's gravity is quite a bit larger than Earth's, we need not be surprised that such a huge equatorial bulge is pushed out.

Although Jupiter has the shortest period of rotation and the fastest equatorial speed, it doesn't have the highest oblateness.

TABLE 18

Period of Rotation of the Planets

PLANET	PERIOD OF ROTATION IN HOURS	IN DAYS
Mercury	1,409	59
Venus	5,834	243
Earth	24	
Mars	24.6	
Jupiter	**9.85**	
Saturn	10.23	
Uranus	10.82	
Neptune	15.7	
Pluto	150	6.4

TABLE 19

Equatorial Speed of the Planets

PLANET	SPEED OF EQUATORIAL SURFACE IN MILES PER HOUR	IN KILOMETERS PER HOUR
Mercury	7	11
Venus	4	6.5
Earth	1,040	1,680
Mars	540	870
Jupiter	**28,000**	**45,000**
Saturn	23,000	37,000
Uranus	8,500	13,700
Neptune	5,600	9,000
Pluto	85	140

Saturn does. We must remember here that Saturn's smaller equatorial speed fights against a gravitational pull that is smaller than that of Jupiter, so that Saturn succeeds in pushing out the greater bulge. (In the same way, Mars has a smaller equatorial speed than Earth has, but also has a smaller gravity, so that it works up a wider bulge in proportion to its size.)

From the oblateness of a planet, and from the direction in which the spots on its surface move, it is possible to tell exactly where its equator is. The equator always stretches across the widest portion of the globe. The axis of rotation runs through the planet at right angles to the equator and across the narrowest portion of the globe.

It might seem a particularly neat arrangement to have a planet rotate so that its axis is perpendicular to the plane in which it revolves about the sun. In that case, the plane of the planet's equator would also be the plane of the planet's revolution. Someone standing on the equator of such a planet would always see the sun at noon directly overhead or at the zenith.

If we imagined the planet as moving about the sun in a horizontal plane, the axis would then be straight up and down, the north pole on top (as we usually picture the Earth's globe, for instance) and the south pole at the bottom.

According to modern theories of the origin of the solar system, such a situation would be easy to explain—but that's not the way things are. The axis of rotation is not perfectly vertical, but is always tipped from the vertical through some particular angle.

In the case of Earth, for instance, the axis of rotation is tipped 23.5° to the vertical.

It is because of this tipping that we have the seasons. The direction of the tilt never changes position as the Earth moves around the sun. At one end of the orbit, the tipping of the axis slants the northern hemisphere more toward the sun and the southern hemisphere more away from the sun. It is then summer in the northern half of the world with the noonday sun high in the sky; winter in the southern half with the noonday sun low. On the other end of the orbit, it is the northern half that is tipped away, the southern half toward, and the situation is reversed. In between the two extremes are spring and fall.

The axial tilt of the various planets is given in Table 20, which, as it happens, contains many puzzles. Why are so many of the planetary axes tipped through an angle between 23° and 29°? Earth, Mars, Saturn, Neptune, and probably Mercury, more than half the total, are included in this group. That axes are tipped at all requires explanation, but why so often by just so much? Then, too, why is the axis of Uranus tipped through an angle of more than 90° so that it seems to be rotating on its side? Why is that of Venus tipped through such a large angle that the planet is standing on its head, so to speak, with its north pole pointing downward and its south pole upward?

TABLE 20

Axial Tilt of the Planets

PLANET	AXIAL TILT (IN DEGREES)
Mercury	28 (?)
Venus	177
Earth	23.5
Mars	25.2
Jupiter	**3.1**
Saturn	26.7
Uranus	97.9
Neptune	28.8
Pluto	?

None of these questions can be answered by astronomers as yet. They can only turn with relief to Jupiter, the only one of the planets with its axis sturdily upright—or at least almost so. Its axial tilt is only 3.1°.

Mass

On several occasions so far I have mentioned gravity in connection with the planets: Jupiter has a stronger gravitational field than Saturn; the sun has a stronger one than Earth.

How do we know this? To answer that question, suppose we think about gravity for a while.

When Newton worked out the mathematical rules governing gravitation, he pointed out that the gravitational effect produced by any body weakens with distance. As one proceeds out into space farther and farther away from Earth, its gravitational effect on any other particular body grows less.

This would show in connection with an object that circles the Earth. An object circling the Earth at a particular distance always moves at a speed that depends on the strength of Earth's gravitational field.

The moon, for instance, which is at an average distance of 239,000 miles from the center of the Earth, moves around the Earth at an average speed of 0.64 miles per second. If it were farther from the Earth it would move more slowly; if it were closer to the Earth it would move more rapidly. If Earth had two bodies circling it, the one farther away would move more slowly than the one nearer to us.

Newton's theory tells us this, but can we be sure? Actually, Earth happens to have only one body circling it—only the Moon and no other. How can we know, then, how other bodies, nearer or farther, would really behave since they are not there to be observed?

What if we consider the sun instead? The sun has nine bodies circling it, the nine planets, at varying distances. The greater the distance, the more slowly the planet moves (see Table 10 on page 46). The manner in which their orbital speeds grow slower as their distance grows greater fits in exactly with Newton's theory.

We now have a chance to study two gravitational effects. We can study the gravitational pull of the Earth on the moon, since we know how far the moon is from the Earth and how fast it circles us. We can also study the gravitational pull of the sun on the Earth, since we know how far the Earth is from the sun and how fast it circles the sun.

Earth is much farther from the sun (93,000,000 miles) than the moon is from the Earth (239,000 miles), and yet the sun's gravitational pull on Earth is much stronger than Earth's pull on the moon. The Earth manages to force the moon to travel at only 0.64 miles per second, but the sun forces the Earth to travel at 18.5 miles per second, even though the sun is so far away.

Allowing for the differences in distance, we can use Newton's mathematical equations to show that the difference in orbital speeds of the moon and Earth means that the sun's gravitational field is 333,400 times as strong as that of the Earth.

By Newton's theory, the strength of the gravitational field of any body depends on something called its "mass," so we can say that the sun's mass is 333,400 times as much as the Earth's.

The easiest way of picturing what mass is in a general kind of way (and under ordinary circumstances) is to say that it varies with the amount of matter in any object. An elephant is made up of more matter than a man, and a man is made up of more matter than a mouse. We can expect, therefore, that an elephant would have more mass than a man or to be more "massive" than a man. In the same way, a man would be more massive than a mouse.

In general, you can expect large things to be more massive

than small things. You could expect objects with a large volume to have a large mass also, since they have more room to pack matter into.

The strength of the pull of the Earth upon different objects depends on the mass of those objects as well as on the mass of the Earth. An elephant would be pulled more strongly by the Earth than a man would. A man would be pulled more strongly than a mouse would.

To measure the pull of the Earth on a particular object, we weigh the object. If we stay on the surface of the Earth (so that the Earth's gravitational pull will stay the same), we can get an idea of the mass of an object by measuring its weight.

For that reason, mass is measured in the same units as weight. In the United States, the ordinary units used for measuring weight are ounces and pounds, with one pound equal to sixteen ounces. In the metric system, used in the rest of the world and by all scientists, even in the United States, the unit of weight is the gram, which is equal to about $\frac{1}{27}$ of an ounce. A thousand grams is equal to a kilogram, and that is equal to 2.2 pounds.

Density

Once we start weighing things, however, we find out that we can't always be sure of mass just by noticing how large an object is. A solid brick, the kind that goes into constructing a wall, weighs a certain amount. A solid wooden brick of the same size would weigh considerably less, and a solid iron brick of the same size would weigh considerably more.

What counts in some ways, then, is not just the size, but how much mass is squeezed into a particular volume. This is called "density."

For instance, the volume of the sun is 1,400,000 times that of the Earth, but the mass of the sun is only 333,400 times that of the Earth. If the material of the sun were exactly like that of the Earth, you would expect the volume of the sun to contain 1,400,000 times the mass of the sun. In actual fact, the volume of the sun contains only 0.24 times that expected mass. The density of the sun is only 0.24 that of the Earth.

This is perhaps not surprising. Scientists have noticed that when any substance is heated up, it usually expands. Its mass takes up more volume, so that its density drops. The sun is much, much hotter than the Earth, so one would expect, for that reason alone, that its density might be lower. Then, too, the sun may be made up of materials that are different from and less dense in general than the materials making up the Earth. This is something we will come back to later in the book.

So far we have been comparing the mass and density of the sun to that of the Earth, but what are the actual figures? What are the actual mass and density of the Earth? If we knew that, we would also work out the mass and density of the sun.

Determining the actual mass of the Earth is difficult. The obvious way would be to place the Earth on a pair of scales and weigh it, but that can't be done.

If we put that to one side for a bit, however, we might consider density. How is that measured? The density of any object is equal to the mass a particular volume of it contains. In the United States a volume is measured as so many cubic inches, where each cubic inch can be pictured as a small cube that is one inch on each side. We can therefore say that the density of something is equal to so many ounces per cubic inch.

In the metric system, however, volume is measured in cubic centimeters. A cubic centimeter is a cube that is one centimeter long on each side. A centimeter is equal to $\frac{2}{5}$ of an inch, and a cubic centimeter is equal to about $\frac{1}{16}$ of a cubic inch. In the metric system, density is measured as so many grams per cubic centimeter.

Well, then, can we take some chunk of the Earth, weigh it and measure its volume, and then calculate its density?

Yes, we can, but that would only give us the density of that particular chunk. It would not give us the density of the entire Earth, unless the Earth were made up of the same materials as that chunk under the same conditions, which is not the case.

There is a great deal of water on the surface of the Earth, and a great deal of air. The density of water is 1 gram per cubic centimeter, and the density of air near the surface of the Earth is 0.0013 grams per cubic centimeter. This is not representative of the density of the Earth as a whole, for the rocks that make up the Earth's solid substance are denser. A cubic centimeter of rock surely has a mass of greater than 1 gram.

In fact, if we take samples of all kinds of different rocks here and there on the Earth and average their densities, we find that the surface rocks of the Earth have an average density of 2.8 grams per cubic centimeter.

Is that, then, the density of the Earth as a whole? Very likely not, for as we imagine ourselves considering rocks deep down in the Earth—ten miles down, a hundred miles down, a thousand miles down—we can see that those deep-lying rocks must be compressed by the weight of the rock above. The rocks deep down are squeezed together so that their mass takes up less room. Their densities are therefore higher.

What's more, when the Earth formed, very dense material may have sunk to the center, while less dense material was left floating on top. In that case, the densities of the rocks on the surface would not be a fair example of the density of the Earth as a whole. The surface density would be less than the over-all density.

This is surely so. In fact, from the way earthquake shocks behave, scientists are pretty certain that the center of the Earth is made up of a large core of liquid iron, which is much denser than the surface rocks.

The only way, then, to determine the average density of the Earth is to determine the total mass of the Earth and its total volume and then divide the mass by the volume. Determining the volume is easy once we know the Earth's diameter, but we are still left with the problem of determining its mass.

Perhaps we can compare the strength of its gravitational field to that of some small object. The Earth's gravitational field would be greater by a particular amount than that of the small object, and the Earth's mass would be greater by that same amount than the mass of the small object. We can then weigh the small object. From its weight we know its mass, and from its mass we know the Earth's mass.

This was done in 1798, by an English scientist named Henry Cavendish. He measured the very tiny pull on a small metal ball by the gravitational influence of a large metal ball. In this way, it turned out that the Earth's mass is just about six billion billion billion grams (6,000,000,000,000,000,000,000,000,000 grams). The sun's mass is 333,400 times that.

It is good to know the Earth's mass, but with all those zeroes it is easier to continue to describe the mass of other planetary bodies as so many times that of the Earth.

Once we know Earth's mass and volume, we can determine its density. That turns out to be 5.52 grams per cubic centimeter. The sun's density is easily calculated to be 1.41 grams per cubic centimeter.

But now that we have gone through this, we must ask the question that particularly interests us because of the subject of this book. What is the mass of Jupiter, and what is its density?

We worked out the mass and density of the sun because we could compare its pull on Earth to Earth's pull on the moon. Can we now compare Jupiter's pull on Earth to Earth's pull on the moon?

Unfortunately not. Jupiter has a gravitational field, but it is far smaller than that of the sun, and Jupiter is farther from us than the sun is. Jupiter's gravitational field is so small out here in the neighborhood of the Earth and the moon that it is almost impossible to measure accurately enough to allow us to calculate its mass and density.

If there were bodies close to Jupiter, as our moon is close to Earth, then we could measure the effect of Jupiter's gravitation upon them and get our answer.

As it happens, there *are* bodies closer to Jupiter, and that brings us back to Galileo and his first look at Jupiter through his telescope. . . .

4 ◉

THE
SATELLITES
OF
JUPITER

Planetary Companions

On January 7, 1610, Galileo looked at Jupiter with his telescope and almost at once noticed three little sparks of light near it—two on one side, one on the other, all in a straight line. Night after night he returned to Jupiter, and always those three little bodies were there. Their positions changed. Sometimes they were a little closer to Jupiter, sometimes a little farther, sometimes on one side, sometimes on the other, but they never moved away from Jupiter altogether. On January 13 he noticed a fourth object.

He came to the conclusion that there were four small bodies circling Jupiter, in just the way our moon circles the Earth.

Galileo's discovery was instantly seen as important for several reasons. For one thing, it was the first time in history that members of the solar system had been discovered that were not visible to the unaided eye.

Then, too, in Galileo's time there was still some doubt as to whether Copernicus's notion that the planets circle the sun was really correct. Some conservatives still held to the notion that all objects in the solar system circle the Earth. Here, at least, were four bodies that clearly circled some body other than Earth. They circled Jupiter.

Kepler, when he heard of this discovery, called the new

bodies "satellites." This is a Latin word for individuals who cluster about some rich and powerful person in hope of being invited to dinner or of receiving gifts. In the same way, the new little worlds seemed to hover about Jupiter.

The term stuck, and we now speak of the moon as the satellite of Earth. Sometimes we call satellites "moons" and speak of the "moons of Jupiter," for instance. This, however, is not good practice. The word "moon" should be reserved for our own satellite. (After all, when we send objects into orbit about the Earth we call them "artificial satellites." We don't call them "artificial moons.")

Galileo who, at that time, was being supported by Cosimo de Medici, Grand Duke of Tuscany, suggested that the new bodies be called the "Medicean planets" in Cosimo's honor. Fortunately, this didn't stick.

Instead, there are a variety of ways of referring to them. The four bodies are sometimes lumped together as the "Galilean satellites," which at least honors the right man. They are also numbered one through four, starting at the one closest to Jupiter. They can thus be referred to as J-I, J-II, J-III, and J-IV. (This system is sometimes applied to other satellites as well, so that Earth's moon can be called E-I. In Earth's case, of course, there is no E-II.)

Most commonly, though, the satellites of Jupiter are known by names drawn from the Greek myths. The Dutch astronomer Simon Marius, shortly after Galileo's discovery, chose names from among those who were associated with Zeus (Jupiter) in the myths. His names are still used.

The innermost of the Galilean satellites is called Io, the second Europa, and the fourth Callisto, after nymphs with whom Zeus fell in love at one time or another. The third is named Ganymede, for a handsome young man whom Zeus carried off to the heavens to serve as cupbearer to the gods.

Although the Galilean satellites were the first to be discovered for any planet other than Earth, they were by no

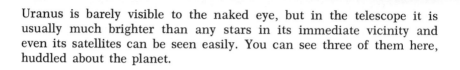

Uranus is barely visible to the naked eye, but in the telescope it is usually much brighter than any stars in its immediate vicinity and even its satellites can be seen easily. You can see three of them here, huddled about the planet.

means the last. As time went on, satellites were discovered for the various other planets. At the present time, thirty-two satellites are known.

Some of the satellites are quite large, two or three thousand miles in diameter. The moon is such a large satellite. So are the Galilean satellites, as we shall see. There are seven such large satellites; the remaining twenty-five are small satellites with diameters ranging from just under a thousand miles to a mere ten miles.

It is quite possible that several more small satellites yet remain to be discovered. The most recent discovery, a satellite of Saturn, was made in 1966. The number of large satellites, however, is surely complete. There is no chance, it would seem, that any large satellite remains to be discovered for any of the known planets.

Table 21 gives the number of satellites for each of the planets. Here you see another difference between the Jovian planets and the Terrestrial planets. The Jovian planets have many satellites—twenty-nine of the total, leaving only three for the Terrestrial planets. And of those three, the two satellites of Mars are tiny bodies only a few miles across.

The Jovian planets have six of the seven large satellites.

TABLE 21

Satellites of the Planets

	SATELLITES	
PLANET	LARGE	TOTAL
Mercury	0	0
Venus	0	0
Earth	1	1
Mars	0	2
Jupiter	**4**	**12**
Saturn	1	10
Uranus	0	5
Neptune	1	2
Pluto	0	0
	7	32

Only one large satellite, the moon, circles a Terrestrial planet (our own), and astronomers are indeed sadly puzzled at the fact. Many theories, none very satisfactory, have been advanced to account for the moon's existence.

It is not surprising, perhaps, that large planets, with large gravitational fields, should have more and larger satellites trapped in those gravitational fields than small planets would. It is also not surprising that Jupiter, the giant of the solar system, should have more satellites than any other and more large satellites than any other.

Indeed, Jupiter is the only planet with more than one large satellite.

Planetary Mass

As soon as the Galilean satellites were discovered, it became possible to determine their period of revolution. One measured the time it took for each to swing from one side of Jupiter to the other and back. This was the time it took each to move in its orbit about Jupiter.

Another thing that could be measured was the amount of separation of each satellite from the center of Jupiter at the extreme of its swing. In Table 22 you will find the period of revolution, and the maximum separation (as measured at Jupiter's closest approach), for each of the Galilean satellites.

Callisto, the farthest from Jupiter of the four satellites, moves from side to side over a stretch of more than 17'. This stretch is more than half the width of the moon as seen from Earth, so Jupiter and its system of satellites cover a surprisingly large stretch of the sky.

Of Uranus's five moons, the fifth to be discovered, Miranda, was the smallest, dimmest, and closest. You see it here just to the right of the planet, inside the four-cornered ring of light which represents the kind of overexposure of the bright central planet that you have to have to get the small Miranda to show up at all.

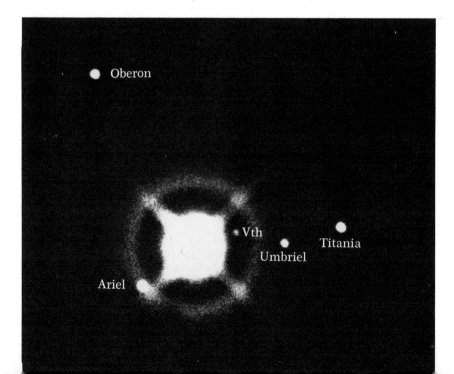

TABLE 22

Revolution and Separation of the Galilean Satellites

SATELLITE		PERIOD OF REVOLUTION (DAYS)	MAXIMUM SEPARATION (MINUTES OF ARC)
J-I	Io	1.77	1.94
J-II	Europa	3.55	3.10
J-III	Ganymede	7.16	4.94
J-IV	Callisto	16.69	8.66

The Galilean satellites all move about Jupiter in less time than it takes the moon to travel about the Earth. Even Callisto, the Galilean satellite farthest from Jupiter, makes its revolution in little over half the time of the moon's 27.32-day period.

Is this because Jupiter's satellites are much closer to Jupiter than the moon is to Earth, so that the former can complete their shorter orbits in a shorter time? Or is it because Jupiter's gravitational force is much greater than Earth's and therefore whips its satellites about faster?

The only way of telling which alternative is correct is to learn Jupiter's actual distance from us. The angular separation of the satellites can then be converted into miles and their speed of revolution determined.

Eventually, as described earlier in the book, the distance of Jupiter *was* determined, and at once the distance of each of the Galilean satellites could be determined. From that, the length of the orbit and the orbital speed could be calculated. The distance and speed for each satellite are given in Table 23. The moon's distance from Earth and its orbital speed are included in this table, for comparsion. (The term "primary" is

Neptune had scarcely been discovered when its larger satellite was seen. The satellite, Triton, as you see, is no brighter than the nearby stars, but its nature is easily understood since it is always found to be very close to Neptune, sometimes on one side, sometimes on the other.

used in the table. This is the name given to the planet around which a satellite revolves.)

As you can see, the four satellites of Jupiter behave just as the planets do with respect to the sun. The farther away a satellite is from Jupiter, the more slowly it moves in orbit. This fits in exactly with Newton's theory of gravitation.

The important point in Table 23 is the manner in which Jupiter's satellites move in orbit so much more quickly than Earth's moon does. Even Callisto, which is five times as far from its primary as the moon is, is nevertheless whipped through its orbit at more than twelve times the speed the moon is.

TABLE 23

Distance and Speed of the Galilean Satellites

SATELLITE		DISTANCE FROM PRIMARY		ORBITAL SPEED	
		MILES	KILOMETERS	MILES PER SECOND	KILOMETERS PER SECOND
J-I	Io	239,000	384,000	10.8	17.4
J-II	Europa	417,000	671,000	8.6	13.8
J-III	Ganymede	665,000	1,070,000	6.8	10.9
J-IV	Callisto	1,170,000	1,880,000	5.0	8.0
E-1	Moon	239,000	384,000	0.64	1.03

This means that there can only be one reason for the swift flight of the Galilean satellites. It is not their closeness to their primary, for they are not particularly close. It can only be the great intensity of Jupiter's gravitational field and, therefore, of Jupiter's great mass.

If we could see Jupiter's satellites with their orbits exactly edge-on, the four of them would form just about a straight line. We see them at a slight angle, so that the orbits look like narrow ellipses and some appear a little higher than others, depending on where each is in its orbit.

August 27, 1916
12ʰ 50ᵐ UT

August 27, 1916
15ʰ 33ᵐ UT

Sept. 4, 1916
12ʰ 50ᵐ UT

From the distance and the orbital speed of the satellites, it is possible to calculate how intense Jupiter's gravitational field is compared to Earth and, therefore, how massive Jupiter is compared to Earth. The satellites of other planets can be used to calculate the masses of those planets in similar fashion. For those planets (Venus, Mercury, and Pluto) which have no known satellites, it is more difficult to determine mass, and more delicate gravitational effects must be used.

In Table 24 the masses of the planets are given.

In terms of mass, Jupiter is more the giant planet of the solar system than ever. It is more than twice as massive as all the other planets put together. In fact, of all the mass in the solar system outside the sun itself, about 70 percent is to be found in a single planet, Jupiter.

A visitor from outer space, looking at the solar system from a distance, might say, "The only object that counts there is this large planet which is fifth from the sun. Everything else makes up just scraps of matter we need not bother with."

TABLE 24

Mass of the Planets

PLANET	MASS (EARTH = 1)
Mercury	0.05
Venus	0.82
Earth	1.00
Mars	0.11
Jupiter	**318.4**
Saturn	95.2
Uranus	14.7
Neptune	17.3
Pluto	0.1

And yet giant Jupiter is still a pygmy compared to the sun. The mass of the sun is 1040 times that of Jupiter. That means that if we consider all the mass of the solar system, *all* of it, 99.9 percent is to be found in that hot, glowing ball, the sun.

Jupiter's mass is surprisingly small in another way, too. Jupiter's volume is 1400 times that of the Earth (see Table 14 on page 52), yet the mass is only 318.4 times as great as that of Earth. Jupiter's mass is spread out more thinly, therefore, over a greater proportionate volume, than Earth's mass is. This means that Jupiter's density is considerably less than that of Earth.

The density of the planets is given in Table 25. As you see, Jupiter's density is only a quarter that of the Earth. In fact, this is another property that sets the Jovian planets apart from the Terrestrial planets: their density is considerably less. Uranus and Neptune also have densities roughly a quarter that of the Earth. Saturn is less dense still, and is actually less dense than water.

This low density of the Jovian planets is something we will come back to later in the book, but for the moment we will stick to the Galilean satellites.

Satellite Sizes

Once we know the distance of Jupiter, we can calculate the diameter of each of the Galilean satellites from its apparent size in a telescope of known magnification.

In Table 26 the diameter of each of Jupiter's satellites is given, along with those of the other three large satellites: the moon, which circles Earth; Titan (S-VI), which circles Saturn; and Triton (N-I), which circles Neptune.

Three of the Galilean satellites, as Table 26 shows, are larger than the moon. So are Titan and Triton. Of the seven large satellites, the moon is, in fact, next to the smallest. Only Europa, just under 2000 miles in diameter, saves it from being at the bottom of the list.

TABLE 25

Density of the Planets

| PLANET | DENSITY | |
	GRAMS PER CUBIC CENTIMETERS	EARTH = I
Mercury	5.1	0.92
Venus	4.85	0.88
Earth	5.52	1.00
Mars	3.96	0.72
Jupiter	**1.34**	**0.24**
Saturn	0.71	0.13
Uranus	1.27	0.23
Neptune	1.58	0.29
Pluto	4	0.7

TABLE 26

Diameter of the Large Satellites

| SATELLITE | | DIAMETER | |
		MILES	KILOMETERS
J-I	Io	2280	3660
J-II	Europa	1930	3100
J-III	Ganymede	3490	5600
J-IV	Callisto	3150	5050
E-I	Moon	2160	3470
S-VI	Titan	3000	4800
N-I	Triton	2300	3700

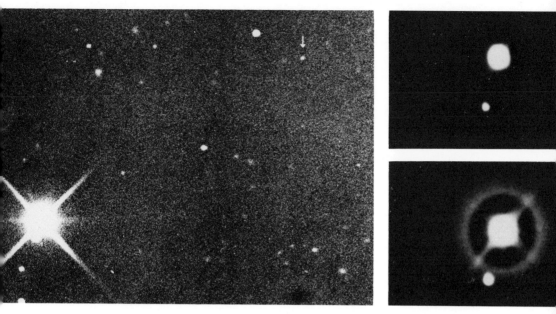

Triton, Neptune's large satellite, can be easily seen, even at Neptune's great distance, with magnifications that don't bring out the nearby dim stars (see right). On the left is Neptune under much greater magnification. Tiny Nereid, Neptune's other satellite, shows up at a considerable distance, dimmer than the many stars. No wonder it was not detected for a century after Neptune's discovery.

Ganymede is not only the largest of the Galilean satellites, it is the largest of all the satellites in the solar system. Callisto is second in both categories. Ganymede and Callisto, for all that they are merely satellites, are each considerably larger than the planet Mercury.

Suppose we consider the surface area of the satellites. The surface area of the moon is about 14,700,000 square miles, or just about as much area as the continents of Africa and Europe combined. How does this compare with the surface area of the other large satellites? The data is given in Table 27.

In this respect, Ganymede is even more predominant. Its surface area is two and a half times the surface area of the moon. Its surface area of 38,400,000 square miles is equal to that of Asia, Africa, and North America combined.

In fact, if we imagine astronauts trying to explore the surface of all four Galilean satellites of Jupiter, they would have to deal with a total surface area of nearly 100,000,000 square miles. This would be six times the surface area of the moon and about half the total surface area of the Earth.

We might also calculate the volume of the large satellites. Here it would be inconvenient to work out the number of cubic miles in each. It would be sufficient for our purposes to calculate the volume of each large satellite on a scale in which the moon's volume is equal to 1. The volumes of the large satellites are given in Table 28.

Ganymede has more than four times the volume of the moon. (If it were hollow, in other words, four moons could be squeezed into it.) The four Galileans together have over nine times the volume of the moon. There is no question but that Jupiter is far more richly supplied with satellites than Earth is in both number and size; far more, in fact, than any of the other planets.

TABLE 27

Surface Area of the Large Satellites

		SURFACE AREA		
SATELLITE		IN SQUARE MILES	IN SQUARE KILOMETERS	MOON = 1.0
J-I	Io	16,200,000	41,800,000	1.1
J-II	Europa	11,700,000	30,200,000	0.8
J-III	Ganymede	38,400,000	99,000,000	2.6
J-IV	Callisto	31,200,000	80,500,000	2.1
E-I	Moon	14,700,000	38,000,000	1.0
S-VI	Titan	28,400,000	73,200,000	1.9
N-1	Triton	16,600,000	42,900,000	1.1

TABLE 28

Volume of the Large Satellites

SATELLITE		VOLUME (MOON = 1)
J-I	Io	1.2
J-II	Europa	0.7
J-III	Ganymede	4.2
J-IV	Callisto	3.1
E-I	Moon	1.0
S-VI	Titan	2.6
N-I	Triton	1.2

What about the masses of the Galilean satellites? These, unfortunately, are difficult to determine.

We know the mass of our own moon quite accurately because its gravitational field has a measurable effect on Earth. In this way, we can tell that the Earth is 81 times as massive as the moon. On the other hand, the Earth is 51 times as voluminous as the moon.

This means that in the body of the planet Earth, 81 moon-masses are squeezed into only 51 moon-volumes. The Earth must be $81/51$ times as dense as the moon. Since the Earth has an average density of 5.52 grams per cubic centimeter, the moon has one of about 3.4 grams per cubic centimeter.

We cannot do for the Galilean satellites what we can do for the moon. Jupiter is larger than the Earth by far, and is less affected by the gravitational fields of its satellites. We cannot measure any influence of the satellites on Jupiter, but we can measure the small influences of one upon another. The determination of mass and density in this fashion is, however, far less precise for the Galilean satellites than for the moon.

Keeping this in mind, the figures for mass and density are nevertheless given in Table 29, with those for the moon also included for comparison.

Of the Galilean satellites, Io and Europa, the two closest, are nearly as dense as the moon and may, on the whole, be made of similar materials. Ganymede and Callisto, the outer two, are distinctly less dense than the moon and are more nearly like Jupiter in density. They must be quite different from the moon in over-all composition.

Another way of comparing the Galilean satellites with the moon is to try to obtain their albedos, the percentage of visible light that they reflect.

TABLE 29

Mass and Density of the Galilean Satellites

	MASS		DENSITY	
SATELLITE	IN TRILLIONS OF TRILLIONS OF GRAMS	MOON = I	IN GRAMS PER CUBIC CENTIMETER	MOON = I
J-I Io	72.5	1.0	2.6	0.8
J-II Europa	47.2	0.6	3.0	0.9
J-III Ganymede	155	2.1	1.6	0.5
J-IV Callisto	97	1.3	1.3	0.4
E-I Moon	73.5	1.0	3.4	1.0

In order to determine the albedo of the Galilean satellites, we must measure their actual magnitude and then compare it with what the magnitude ought to be if the satellites reflected all the light they received.

In Table 30 the magnitude of the Galilean satellites and

their albedo is given. Again the moon is included for comparison.

The most startling aspect of the data in Table 30 is the brightness of the Galilean satellites. Each one of them is bright enough to be visible to the naked eye if it were alone in space. Jupiter, however, is 250 times as bright as even Ganymede, the brightest of the satellites, and the satellites are so close to the planet that its glare drowns them out—at least to the unaided eye.

Io, Europa, and Ganymede all reflect light considerably more efficiently than the moon does, which means that their surface structure must in some way be considerably different from that of the moon. Callisto, on the other hand, is much closer in albedo to the moon.

There is still one point we ought to consider concerning the mass of the Galilean satellites. How great is the mass of each satellite compared to that of the giant planet it circles? Clearly, it must be small indeed. In Table 31, the mass of each satellite is given as a fraction of Jupiter's mass. For comparison, the other large satellites are included, each with its mass expressed as a fraction of that of its primary.

TABLE 30

Magnitude and Albedo of the Galilean Satellites

SATELLITE		MAXIMUM MAGNITUDE	ALBEDO
J-I	Io	4.8	0.37
J-II	Europa	5.2	0.38
J-III	Ganymede	4.5	0.21
J-IV	Callisto	5.5	0.09
E-I	Moon	−12.6	0.06

The startling thing about the data in Table 31 doesn't involve the Galilean satellites at all. They are tiny in comparison to Jupiter, as expected. Even Ganymede has only $\frac{1}{12,000}$ the mass of Jupiter. Titan does somewhat better with respect to its primary, Saturn; and Triton still better with respect to Neptune; but look at the moon!

TABLE 31

Mass of the Large Satellites in Relation to Primary

		MASS IN RELATION TO PRIMARY	
SATELITE		AS FRACTION	AS DECIMAL
J-I	Io	1/26,000	0.000038
J-II	Europa	1/40,000	0.000025
J-III	Ganymede	1/12,000	0.000081
J-IV	Callisto	1/20,000	0.000050
S-VI	Titan	1/4,100	0.00024
N-I	Triton	1/760	0.0013
E-I	Moon	1/81	0.0124

It is $\frac{1}{81}$ the mass of its primary, the Earth. No other satellite in the solar system is anywhere nearly as massive compared to its primary. The large Galilean satellites pale into insignificance compared to Jupiter. All four together are less than $\frac{1}{5000}$ the mass of Jupiter.

In this respect, at least, Earth is the most satellite-rich of all the planets in the solar system. The Earth-moon combination is even referred to as a double planet sometimes—surely the only one in the solar system.

The Fifth Satellite

From the time Galileo announced his discovery of Jupiter's satellites, about forty-five years passed before any new bodies belonging to the solar system were discovered. There remained only five satellites known, the moon and the four Galilean satellites.

Then, in 1655, the Dutch astronomer Christian Huygens discovered a satellite circling Saturn. It was about as far from Saturn as Ganymede was from Jupiter, and about as large as the Galilean satellites. It was not discovered as quickly or as easily as the Galileans because Saturn and its satellites are twice as far from us as Jupiter and its satellites, and are correspondingly dimmer.

Huygens named the new satellite "Titan" because Cronos (Saturn) and his associates were grouped together in the Greek myths as Titans.

This meant six satellites were now known in the solar system and all were 2000 miles or more in diameter. Were all satellites to be that large? On the basis of the first six known, it might have seemed so. But wait. . . .

The first satellites that astronomers would find would naturally be the largest ones, because they are likely to be brightest and therefore the easiest to see. As telescopes were improved, smaller and dimmer bodies could be seen. In the 1670s and 1680s, Cassini discovered four more satellites circling Saturn. These new satellites were distinctly smaller than the moon. All were merely a few hundred miles in diameter.

In fact, of all the satellites discovered since Huygens's discovery of Titan, only one more has proved to be moon-size. This was Triton, a satellite of Neptune (and named for a son of Poseidon, who was known as Neptune to the Romans). Triton was discovered in 1846 by the English astronomer William Lassell, less than a year after Neptune had itself been discovered. (Some of the properties of Titan and Triton have been given in Tables 26, 27, 28, and 31.)

In one respect the situation was ironical as far as Jupiter was concerned. After Galileo's initial discovery, satellite after satellite was discovered for planet after planet—but not for Jupiter.

In the 1600s five satellites of Saturn were discovered. In the 1700s two more satellites of Saturn were discovered, and two satellites of Uranus. In the 1800s two satellites of Mars were discovered, another satellite of Saturn, two more of Uranus, and one of Neptune.

By 1892 fifteen discoveries had been made since Galileo's, and twenty satellites were known altogether: one for Earth, one for Neptune, two for Mars, four for Uranus, eight for Saturn—and there were still only four for Jupiter.

No fifth satellite had been discovered in nearly three centuries of looking. Had Galileo discovered all the satellites Jupiter owned?

In that year of 1892, however, the American astronomer Edward Emerson Barnard began a program of investigating the near neighborhood of Jupiter with a powerful 36-inch telescope. It was a much more powerful telescope than the poor little spyglass with which Galileo had found the four large satellites of Jupiter.

On September 9 Barnard noticed a little speck of light near Jupiter. It was so close to Jupiter and so dim that seeing it in the glare of Jupiter-light was almost impossible. Some have considered the discovery to be the most remarkable one to be made by the human eye in the heavens. (Since 1892, all new satellites to be discovered have been discovered in photographs.)

Barnard followed it carefully and was finally able to announce that he had at last discovered a new satellite of Jupiter.

This new satellite was first called "Barnard's satellite," but was more commonly called "Jupiter-V" or "J-V" because it was the fifth satellite of Jupiter to be discovered.

Jupiter-V has not yet been officially named after some mythological character. The French astronomer Camille Flammarion suggested, about a hundred years ago, that it be named Amalthea. Amalthea was the name of the nymph (or goat) who, in the Greek myths, supplied the milk for the infant Zeus (or Jupiter). It seems rather fitting that though the large satellites be named for the companions of Jupiter's mature years, the small satellite nearest to him be named for the nurse of his childhood. Although the name is not yet official, it will be used in this book.

TABLE 32

Properties of Amalthea

Period of revolution (days)	0.50
(hours)	11.95
Maximum separation (minutes of arc)	0.84
(seconds of arc)	50.5
Distance from Jupiter's center (miles)	112,000
(kilometers)	181,300
Orbital speed (miles/second)	17.2
(kilometers/second)	27.8
Diameter (miles)	70
(kilometers)	110
Surface area (square miles)	15,400
(square kilometers)	38,200
Maximum magnitude	13

Amalthea is not one of the Galilean satellites; that name is applied only to the four large satellites Galileo discovered. In Table 32, some of the properties already given for the Galilean satellites are given for Amalthea.

Amalthea is only about 112,000 miles from the center of Jupiter. Since the surface of Jupiter is 44,000 miles from the center, this means that Jupiter-V circles Jupiter only 68,000 miles above its surface.

Because Amalthea circles Jupiter so closely, it is never seen far from the planet. Even when it is farthest to one side, it is separated from Jupiter's center (as seen from Earth) by an angular distance of only 50.5 seconds of arc when Jupiter is closest. From Jupiter's gleaming edge, Amalthea is separated, at most, by merely 30 seconds of arc. This separation is less than the width of Jupiter's globe.

Even at its greatest separation, Amalthea can barely be seen, and it revolves about Jupiter so quickly (just under twelve hours) that it stays near its greatest distance only an hour or so at a time.

Remember, too, that it is a tiny object compared to the Galilean satellites and is of only the thirteenth magnitude. Io, which is the closest of the Galilean satellites to Jupiter, is nearly three times as far from Jupiter's edge as Amalthea is and is 5500 times as bright. That is why Galileo could spy Io at once with his tiny telescope, while Barnard's discovery nearly three centuries later, with a much better telescope, is so remarkable.

Amalthea's greatest difference from the Galilean satellites is its size. It is only about 70 miles across. Still, its surface area (about 15,400 square miles) is as large as that of Denmark and three times as large as that of the state of Connecticut.

Amalthea was by no means the last of the satellites of Jupiter to be discovered. In the twentieth century, seven more satellites of the giant planet were detected, bringing the total number of known satellites of Jupiter to 12. These seven satellites are a special case, however, and we'll get to them later.

The Speed of Light

The four Galilean satellites, and Amalthea, too, all revolve in very nearly the plane of Jupiter's equator, and in very nearly a circular orbit. The angular difference of the plane of revolution of each satellite from the plane of Jupiter's equator and the eccentricity of each satellite orbit are given in Table 33.

This regularity of satellite orbit—little, if any, tipping from the plane of the equator of the primary, and little, if any, eccentricity—is what we would expect if satellite systems originate in the way most astronomers think they do. The satellites of Uranus all revolve in nearly circular orbits in the plane of the planet's equator. So do the satellites of Mars and eight of the satellites of Saturn.

It is not an invariable rule, though. Our own satellite, the moon, has an orbit that is tipped by 18° to the plane of Earth's equator. The moon's orbit has an eccentricity of 0.05, three times that of Earth's own orbit. These two facts are additional puzzles that astronomers have not yet solved about our satellite.

The regularity of the orbits of Jupiter's satellites is such that if we saw Jupiter with its axis directly up and down, we would see the orbits of the satellites just about on edge. Each satellite would seem to move from west to east in a straight line and then from east to west in the same straight line, and it would recede from Jupiter just as far in one direction as the other.

As each satellite moved from west to east, it would cut in front of Jupiter, moving along the line of the equator. Then, as each moved in the opposite direction, from east to west, it would pass behind Jupiter and be "eclipsed," emerging from the eclipse after a while. Each satellite would be eclipsed in this fashion each time it moved from east to west.

However, from our position on Earth we don't see the satellite orbits quite edge-on. Jupiter's orbit is tipped to that of ours by 1.3° and Jupiter's equator is tipped to its orbit by 3.1°. The

TABLE 33

Orbits of Jupiter's Five Inner Satellites

SATELLITE		INCLINATION (MINUTES OF ARC)	ECCENTRICITY
J-V	Amalthea	3	0.0028
J-I	Io	3	0.0000
J-II	Europa	1	0.0003
J-III	Ganymede	2	0.0015
J-IV	Callisto	21	0.0075

combination of tippings means that from Earth we can sometimes look slightly down on the orbits of Jupiter's satellites from above, and sometimes look slightly up on those orbits from below.

Callisto is at the greatest distance from Jupiter and has an orbit that is most tipped to Jupiter's equator (even though only by a third of a degree) of any of the five. Because of the tipping of its orbit combined with everything else, it can sometimes be glimpsed just above or just below Jupiter's globe as it passes behind the planet.

Most of the time, though, Callisto moves into eclipse. As for the other Galilean satellites, and Amalthea too, those are eclipsed by Jupiter (as seen from Earth) at every single revolution without exception.

After Galileo discovered the four large satellites of Jupiter, astronomers were very much interested in these eclipses. Allowing for the small tipping of Jupiter's orbit and its equator and of the even smaller tipping of the satellite orbits, it seemed an easy task to predict exactly when each of the satellites would move into eclipse and out again.

In fact, some astronomer of the 1600s wondered if the satellites of Jupiter might not be used as a gigantic clock in the heavens. If one of the satellites moved into eclipse, the exact moment at which it disappeared would represent a certain time and the clocks in the observatory (and anywhere on Earth) could be adjusted to that. In this fashion, we would never have to worry about clocks running fast or slow. They could always be corrected by the heavenly clock of Jupiter's satellites.

Naturally, then, astronomers studied the satellites carefully and recorded the exact time of each eclipse, in order to get the necessary data with which to work out the predictions for the future.

There was trouble, though. No matter how carefully astronomers drew up their eclipse tables for the satellites of Jupiter, they never quite workd. At times the eclipses came a bit later than they were supposed to, and at other times eclipses came a bit sooner. The *average* times were correct, but what good was that? What was the use of a clock that told the correct time on the average but was sometimes a few minutes early and sometimes a few minutes late?

Astronomers could think of no reason why the satellites of Jupiter should sometimes be as much as eight minutes early in going into eclipse and sometimes as much as eight minutes late. There was a sixteen-minute deviation altogether.

In 1675 a Danish astronomer, Olaus Roemer, considered the problem. He noticed that when Jupiter and Earth were on the same side of the sun, the eclipses were always earlier than average. As the Earth raced ahead and moved over to the side of the sun opposite to that of Jupiter, the eclipses came later

Ole Roemer was born in Denmark in 1644. In 1675, after closely studying the motions of Jupiter's satellites, he determined the speed of light for the first time. He died in 1710.

and later. Finally when the Earth rounded the sun and began to approach the side on which Jupiter was, the eclipses came earlier and earlier. When Earth and Jupiter were as close as they could be, the eclipses came about 16 minutes sooner than when Earth and Jupiter were as far apart as they could be.

Suppose, Roemer thought, it took time for light to travel over a particular distance. It might be that we didn't see the Galilean satellites at exactly the moment they were in a particular spot, but only when the light from the satellites reached us later, when they were no longer in that particular spot.

When a satellite passed behind Jupiter, its light was cut off. We on Earth, however, would not see the light cut off and the satellite pass into eclipse until that cutoff point reached us.

When Jupiter and Earth are on the same side of the sun, the average distance between the two is about 390,000,000 miles. When Jupiter and Earth are on opposite sides of the sun, the average distance between the two is about 586,000,000 miles.

When Jupiter and Earth are on opposite sides of the sun, the light from the satellites must travel an additional 186,000,000 miles on the average—the complete width of the Earth's orbit. Suppose it took the light rays 16 minutes to cross that extra distance. That would exactly account for the way in which the eclipses lagged as Earth moved away from Jupiter, and caught up again and moved ahead, as Earth traveled toward Jupiter.

This meant that light would have to travel about a hundred eighty thousand miles each second. Indeed, the speed of light according to modern measurements is 186,282 miles per second.

This is extraordinarily fast by Earthly times. Light travels so quickly that in human affairs, no one need allow for it. If you see something in a certain place, that is where it is, for light takes only the tiniest fraction of a second to pass from the object to your eye, and in that tiny fraction of a second, the object has no time to move away.

This is a far cry from a modern astronomical observatory. It is Roemer's work place, however, and the picture shows the telescope with which he studied the satellites of Jupiter and determined the speed of light.

When we move on to bodies outside the Earth, the speed of light must be taken into consideration. It takes 1.28 seconds for light to travel from the moon to the Earth. The interval of time becomes longer and longer for bodies that are farther and farther from Earth.

The distance of 186,282 miles per second, which is the distance light travels in one second, is called a "light-second." Thus, the average distance of the moon from the Earth is 1.28 light-seconds.

A "light-minute" is a little over 11 million miles, sixty times as long as a light-second, since it is the distance that light will travel in sixty seconds. In the same way, a "light-hour" is nearly 671 million miles, or sixty times that of a light-minute.

In Table 34 the average distance of the planets from the sun is given in light-units. You can see that it takes a little over eight minutes for light to reach us from the sun. It takes well over five *hours*, however, for light to travel from the sun to Pluto. We might say that it would take light nearly 11 hours to travel from one side of Pluto's orbit to the other, so that the extreme diameter of the sun's planetary system would be shown to be nearly 11 light-hours.

This seems like a great deal but it shrinks to nothing, almost, when distances outside the solar system are taken into consideration. In the nineteenth century it was discovered that the stars are so distant that it takes light many *years* to travel from them to Earth. The "light-year" became a common unit of distance to astronomers interested in the stars.

What's more, the speed of light turned out to be one of the most important values known to science. It sets an absolute limit on how fast anything can move and how fast anything can happen. Einstein's famous theory of relativity is based on the speed of light.

And that speed was first measured in 1675 through studies of the eclipses of the Galilean satellites of Jupiter.

TABLE 34

Distance of Planets From the Sun in Light-Units

PLANET	AVERAGE DISTANCE FROM THE SUN	
	IN LIGHT-MINUTES	IN LIGHT-HOURS
Mercury	3.2	0.053
Venus	6.0	0.10
Earth	8.3	0.14
Mars	12.6	0.21
Jupiter	**43.3**	**0.72**
Saturn	79.4	1.32
Uranus	160	2.68
Neptune	250	4.16
Pluto	320	5.33

5

THE
INFLUENCE
OF
JUPITER

Jupiter and Saturn

As I said earlier in the book, the gravitational field of any object grows weaker in proportion to the square of the distance. It never falls quite to zero, however, no matter how great the distance. Every body in the Universe therefore has an influence over every other body, however far apart the two may be.

Of course if the distance is great enough, and the bodies are small enough, the pull between them becomes so small that it can safely be ignored for all practical purposes. The gravitational pull of the small planet Mercury on the distant planet Pluto can be ignored without affecting any astronomical calculations.

The more massive a body, the farther out into space its influence stretches before it becomes small enough to be ignored. Since Jupiter is by far the most massive planet in the solar system, its gravitational influence is most important. And since Saturn is the next most massive planet and the two are neighbors, the gravitational influence between them is stronger than that between any two other planets.

When Jupiter and Saturn are on the same side of the sun and are as close to each other as possible, they are 328,000,000 miles apart. Saturn is then 280 times as far away from Jupiter as Callisto is. On the other hand, Saturn is 5,700 times as massive as Callisto. It can be calculated that the gravitational attraction between Jupiter and Saturn, when those two planets are at their closest, is about $\frac{1}{14}$ that between Jupiter and Callisto and is actually 1.1 times as strong as the attraction between the Earth and the moon.

This is quite an attraction. If Jupiter and Saturn were separated by that distance and were alone in the universe, they would circle each other about a point some 76,000,000 miles from Jupiter. They would form a double planet, each making a complete turn about that intermediate point in a little over four and a half years.

The reason that they don't do so is that the sun also exists. Although the sun is more than twice as far from Saturn as Jupiter is, the sun's enormous mass makes its gravitational pull on Saturn supreme. The sun's pull on Saturn is about 150 times that of Jupiter's pull on Saturn, even when the two planets are at their closest.

But although Saturn obeys the sun's superior pull and re-volves about it, that doesn't mean it isn't affected somewhat by Jupiter's pull, too. As Jupiter wheels about the sun and ap-proaches Saturn's position (remember that Jupiter moves faster in its orbit about the sun than the more distant Saturn does), the gravitational attraction between them increases. Jupiter's pull, as it catches up with Saturn, slows down the other planet. Saturn's pull, on the other hand, speeds up Jupiter.

After Jupiter passes Saturn, the situation reverses. Now it is Jupiter that is ahead, and its pull speeds up Saturn, while Saturn's pull slows down Jupiter.

In the end, neither planet is permanently affected; the speedups and slowdowns cancel each other. Still, scientists who try to work out systems for calculating exactly where Jupiter and Saturn will be in their orbits must take the pull between them into account.

If it were only necessary to consider the gravitational attrac-

Saturn has been called the most beautiful object we can see in the heavens. Certainly there is nothing else like it. In this photograph we see the rings as clearly as we possibly can from Earth, with the planet's axis tipped as far toward us as it can get.

tion between the sun and a particular planet, the movements of that planet would be very easy to calculate. The smaller pulls between the planets, however, disturb or "perturb" this simple situation and astronomers must calculate the amount of these "perturbations."

Since Jupiter is the most massive planet in the solar system, it is responsible for the largest perturbations on the other planets and does the most to make calculations hard for astronomers.

Naturally, none of the other planets is perturbed by Jupiter to the extent that Saturn is. No other planet except Mars (occasionally) approaches Jupiter as closely as Saturn (occasionally) does.

Suppose we set the attraction between Jupiter and Saturn

at their closest as being equal to 10,000. We can then set numbers, in proportion, for the attraction of Jupiter for any other planet at the point of their closest approach. (On this scale, the attraction between earth and moon is 9,100.) The attraction of Jupiter for the other planets on this scale is given in Table 35.

Clearly, the attraction of Jupiter for the other planets is much less than its attraction for Saturn, and yet the quantities cannot be ignored. When Jupiter is closest to Earth, its pull on Earth is stronger than that of any other planet—twice as

TABLE 35

Pull of Jupiter on Other Planets

PLANET	PULL OF JUPITER (SATURN = 10,000)
Mercury	3.3
Venus	60
Earth	89
Mars	13
Jupiter	——
Saturn	10,000
Uranus	120
Neptune	38
Pluto	0.2

strong as the pull of Venus at its closest, and Venus is our nearest planetary neighbor. In calculating the details of Earth's motion through the solar system, the pull of the sun must be taken first into account, and then the pull of the moon. After that, though, Jupiter comes third.

the most massive of the inner planets, is very little compared to the fact that the distance from the center of the sun to its surface is 432,000 miles.

Distant Pluto is so far away from the sun that even though its mass is no larger than Mars, it manages to revolve about a center of gravity that is 1,200 miles from the sun's center. This still isn't much, however. It is only $\frac{1}{36,000}$ of the way from the center to the surface of the sun.

Only the four Jovian planets—Jupiter, Saturn, Uranus, and Neptune—circle a center of gravity that is quite far from the sun's center. And of these, Jupiter circles about a center of gravity farthest from the sun's center, even though it suffers under the handicap of being the closest of the giant planets.

Indeed, the center of gravity of the sun–Jupiter system is the only one that lies outside the body of the sun. It is 460,000 miles from the center and the sun's surface is only 432,000 miles from the center. The center of gravity of the sun–Jupiter system is therefore 28,000 miles above the sun's surface (in

TABLE 36

Center of Gravity of the Sun-Planet Systems

PLANET	CENTER OF GRAVITY (MILES FROM SUN'S CENTER)
Mercury	6
Venus	160
Earth	300
Mars	50
Jupiter	**460,000**
Saturn	250,000
Uranus	80,000
Neptune	150,000
Pluto	1,200

the direction of Jupiter) and the sun revolves about it in 11.86 years, keeping perfect step with Jupiter.

If Jupiter were the only planet in the solar system, that would be it. Jupiter and the sun would both make a complete circle in 11.86 years, Jupiter a large one and the sun a small one.

Under those conditions, someone looking at the solar system from a great distance might not be able to see Jupiter and yet he would see something else. He would notice the sun seeming to wobble in its position by a very tiny amount, making one complete wobble in 11.86 years. From that, the observer could deduce that there was a large planet circling the sun, even though he couldn't see it. From the exact manner of the shift, he might even be able to deduce something about the mass and distance of the unseen planet.

Of course, the fact that there are other planets in the solar system besides Jupiter complicates things. The sun must actually move about the center of gravity of the entire solar system, a center of gravity to which all the planets contribute. If all the other planets happened to be on the side exactly opposite from Jupiter, their contributions all together would somewhat outweigh that of Jupiter. The center of gravity of the solar system would then be some forty thousand miles from the center of the sun, on the side *away* from Jupiter.

The center of gravity of the solar system shifts in a complicated fashion as the planets circle the sun, but most of the time it points in the general direction of Jupiter.

The Asteroids

What else does Jupiter's gravitational effect accomplish? We have discussed its perturbations on other planets and its effect on the sun, and, of course, there is the fact that it holds its satellites in orbit about itself. Is there anything else?

Look at the solar system again and consider the distances of the planets from the sun.

How much farther away from the sun is a particular planet than the one next closer to the sun? For instance, Venus, with an average distance from the sun of 67 million miles, is 1.9 times as far from the sun as its inner neighbor Mercury, which is an average distance of 36 million miles from the sun.

Table 37 shows the ratios of distance for each pair of planets. The ratio shifts from planet to planet but, except in one case, not by very much. All the ratios but one lie between 1.3 and 2.0. Leaving out that one ratio, we might say that each planet is, roughly, a little over one and a half times as far from the sun as the planet before.

The one exception is the distance between Jupiter and Mars. Jupiter is 3.4 times as far from the sun as Mars is. The gap between those two planets is twice as great as it might be expected to be. It is almost as though a planet ought to exist between Mars and Jupiter, and doesn't.

Or does it? Could there be a planet between Mars and Jupiter that had not been discovered?

It didn't seem likely. It is easy to see Mars and Jupiter in the sky even without a telescope. Could a planet between them go unnoticed right down to the end of the 1700s? The only way in which that could be so would be for such a planet to be very small and therefore very dim.

Toward the end of the 1700s, astronomers were thinking along those lines and were beginning to plan a telescopic sweep of the sky in order to see if such a missing planet could be spotted—any object that moved against the stars more quickly than Jupiter but more slowly than Mars.

On January 1, 1801, an Italian astronomer, Giuseppe Piazzi, detected such a planet and named it Ceres, after the Roman goddess of agriculture. It was a very interesting way to begin the new century.

The orbit of Ceres was calculated by a German mathematician, Carl Friedrich Gauss, and it was found to lie between that of Mars and Jupiter.

TABLE 37

Ratios of Distances of Planets From the Sun

NEIGHBORING PLANETS	AVERAGE DISTANCE RATIO
Venus-Mercury	1.9
Earth-Venus	1.4
Mars-Earth	1.5
Jupiter-Mars	3.4
Saturn-Jupiter	1.8
Uranus-Saturn	2.0
Neptune-Uranus	1.6
Pluto-Neptune	1.3

The average distance of Ceres from the sun is 250 million miles. It is 1.8 times as far from the sun as Mars is, and Jupiter is 1.9 times as far from the sun as Ceres is. If Ceres were included in Table 37, the ratios would be fairly even all the way down the line.

In a way, the discovery of Ceres was embarrassing. Mars had been considered the 4th planet so long, and Jupiter the 5th planet, that it seemed impossible to upset the numbering. Could Ceres be called the 4½th planet? Fortunately, it soon turned out there was a good reason for not giving Ceres a number at all.

Ceres is only 460 miles in diameter, far smaller than any other planet. It has only ⅐ the diameter of Mercury, the smallest planet known up to the discovery of Ceres. Ceres was, in fact, the smallest object to be discovered within the solar system up to that point—which was exactly why it had not been discovered earlier. It was so small, and reflected so little light, that even when it was closest to Earth it was not quite bright enough to be seen with the unaided eye.

Ceres was so small that astronomers wondered if that was all there was in the huge gap between Mars and Jupiter. They kept on looking. In March 1802 a German astronomer, Heinrich Olbers, discovered a second planet in the gap, one with an orbit not far from that of Ceres. It was named Pallas (an alternate name for the Greek goddess Athena).

This was something unprecedented! Two planets with nearly the same orbit! What's more, Pallas was even smaller than Ceres, being only some 300 miles in diameter. In 1804 a third planet, Juno (the wife of Jupiter), was discovered and in 1807 a fourth planet, Vesta (the Roman goddess of the hearth). Both traveled along orbits between those of Mars and Jupiter and both were under 200 miles in diameter.

Although these four worlds, all between Mars and Jupiter, are small, they are planets. They circle the sun independently and therefore deserve the name. Yet because of their small size they are sometimes called "minor planets" or "planetoids."

Their most common name, however, is "asteroid," meaning "starlike." It was given them by Herschel, the discoverer of Uranus. He selected the name because the new little worlds are so small that when they are viewed through a telescope they are not seen as globes, as the other planets are, but as points of light, as stars are.

In fact, the diameters of the asteroids cannot be worked out from the size of their disc, but from the amount of light they reflect. If the distance of the asteroid is known and if it is assumed that it reflects as much light as an airless object like the moon does, then the diameter can be worked out.

It was inconvenient to give each of the four asteroids a number and to have Jupiter the ninth planet from the sun. What if still more asteroids were discovered? Besides, once the small bodies were called asteroids rather than planets, it was easy to skip them in the numbering. Asteroids and planets could be numbered separately: the asteroids in order of discovery, planets in order of distance.

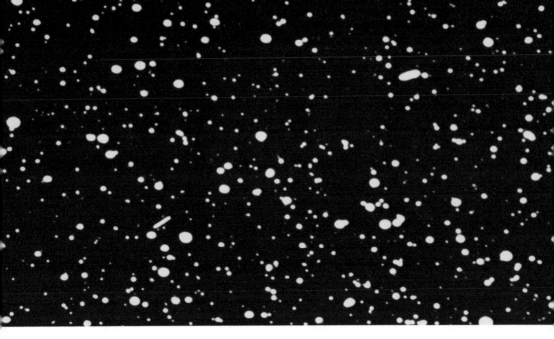

When photographs of the stars are taken, with the camera moving in time to the turning sky, all the stars are dots of light. Something which is moving against the background of the stars, and is no brighter than the stars, is an asteroid. You can see two in this photograph, producing short lines rather than round dots. The fatter line marks the asteroid Bellona, 28th to be discovered.

After 1807, astronomers settled down for a time in the belief that there were four asteroids circling in the "asteroid belt" that lay between Mars and Jupiter. But then, in 1845, a German astronomer, Karl L. Hencke, discovered a fifth, which he named Astraea after the Greek goddess of justice.

After that still others were discovered, and then more and more and more. By now, over 1600 asteroids have been detected and named, and there may be thousands more that are as yet undetected.

Jupiter and the Asteroids
The question is, then, why is there this asteroid belt between the orbits of Mars and Jupiter?

Two answers have been suggested, and both may involve Jupiter.

First, perhaps a planet—a single planet—existed there to begin with. It wouldn't have been a large planet, perhaps no larger than the moon. If at some time it exploded, would the asteroids be the fragments of the explosion?

But why would the planet explode? Might it be the result of Jupiter's gravitational field? None of the large planets approaches much closer than 400 million miles to Jupiter, but the "asteroid-planet" would have approached to within 200 million miles of Jupiter every seven years. Could Jupiter's gravitational effect, exerted every seven years, have made the asteroid-planet unstable enough to give it the nudge that caused it to explode?

This conjecture poses many difficult problems. Most astronomers are quite dubious about the explosion theory.

What about the reverse? Suppose the region between the orbits of Mars and Jupiter began with a large number of fragments that *never pulled together* to form a planet?

The modern theory of the formation of the solar system has its beginning as a huge cloud of dust and gas. Slowly this cloud turned, and slowly it came together under its own gravitational pull.

As the cloud condensed into a smaller and smaller object, it turned faster and faster. Eventually, the central part of it condensed into the sun, while some of it at its midsection was kept in the outskirts by the centrifugal effect, like a huge equatorial bulge. The thinner cloud of dust and gas that spread out beyond the sun's midsection formed larger and larger objects that kept colliding and sticking until the planets were formed, all circling more or less in the equatorial plane of the sun.

Could it be that the part of the dust cloud lying between the orbits of Mars and Jupiter had collected into small solid bodies of various sizes but could not take the final step of coalescing into a single large body?

Why couldn't it? Jupiter again. As Jupiter whirls around the sun, its gravitational influence keeps stirring up the asteroids, and preventing them from slowly coming together. At least so it might seem.

This second theory is rather attractive. It might even explain why Mars is as small as it is.

If we consider the four inner planets, Mercury, the closest to the sun, is the smallest. Venus, the second, is considerably larger than Mercury, and Earth is a little larger than Venus. Why, then, shouldn't Mars be even larger than Earth, instead of being only a ninth as massive?

Perhaps it was the influence of Jupiter. Jupiter is farther from Mars than it is from the asteroids, so it didn't prevent the formation of a planet. Perhaps, though, Jupiter's vast gravitational field kept some of the fragments from getting together to form Mars; perhaps Jupiter's field gathered up many of the fragments between Mars and Jupiter and left unusually few for Mars.

It is perhaps because of Jupiter's vast greed that Mars is only ⅑ as massive as Earth and that all the asteroids put together are less than ⅑ as massive as Mars.

The effect of Jupiter on the asteroids is not confined to the dim and distant past, either. In one way, the effect can be seen clearly right now.

By 1866 enough asteroids had been discovered so that one could see that the average distances were spread out fairly evenly between the orbits of Mars and Jupiter. —But not entirely evenly.

An American astronomer, Daniel Kirkwood, noted that at certain distances from the sun there seemed to be no asteroids. For instance, there were no asteroids with an average distance from the sun of 230 million miles, or 275 million miles or 305 million miles or 340 million miles. Was there a reason for that?

Well, consider the average asteroid.

Obviously, the most important perturbations it receives will be from Jupiter's huge gravitational field. Every time the asteroid wheels into that part of its orbit which happens to be near Jupiter's position at the time, it feels Jupiter's pull particularly strongly.

If Jupiter happens to be a little ahead of the asteroid at the time of closest approach, it will pull the asteroid forward. If Jupiter happens to be a little behind, it will pull the asteroid backward. In general, the asteroid will be pulled forward on some of the close approaches and backward on some of the other close approaches. On the average, the forward and backward pulls will cancel each other and, in the long run, the asteroid's orbit will remain unchanged.

But suppose an asteroid has an orbit which places it at an average distance of 305 million miles from the sun. In that case, it will make one circle round the sun in just under six years. That is just one half of Jupiter's period of revolution.

Suppose such an asteroid approaches the point in its orbit close to Jupiter's when the giant planet is a little ahead. As the asteroid catches up, its elliptical orbit carries it farther away. This means that the strongest pull of Jupiter on the asteroid will be forward. When the asteroid makes two complete revolutions and returns to that point on its orbit, Jupiter has made one complete revolution and is a little ahead again. The two bodies are in the same position they were twelve years before.

In fact, every twelve years the asteroid and Jupiter are in the same positions relative to each other. If Jupiter is a little bit ahead at the time, the asteroid gets yanked forward *each time*. If Jupiter is a little behind, the asteroid gets yanked backward *each time*. There is no balancing of perturbations.

Asteroids that regularly get yanked forward move out farther from the sun. Asteroids that regularly get yanked backward move in closer to the sun. In either case they don't stay where they were. It is for this reason that there are no asteroids with an average distance of 305 million miles. Any asteroid of this type is pushed farther in or farther out by Jupiter.

This is true whenever the period of revolution of an asteroid is some simple fraction of the period of revolution of Jupiter. Any asteroid at an average distance of 230 million miles from the sun has a period of revolution just ⅓ that of Jupiter. Again, every twelve years they are in the same relative positions.

As long ago as 1776, the Italian-French astronomer Joseph Louis Lagrange pointed out that this would take place under the proper conditions. At that time the asteroids had not yet been discovered, and Lagrange knew no actual case in which the sweeping-clean of certain orbits took place. He worked it out as pure theory.

Ninety years after Lagrange's theoretical work, however, Kirkwood showed that it applied perfectly to Jupiter and the asteroids. Those places between Mars where no asteroids would be found because of Jupiter's perturbations have been known as "Kirkwood's gaps" ever since.

The Trojan Asteroids

A still closer connection between Jupiter and the asteroids became clear, beginning in 1906. In that year a German astronomer, Max Wolf, discovered asteroid number 588. It was unusual because it moved at a surprisingly slow speed and therefore had to be surprisingly far from the sun. It was, in fact, the farthest asteroid that had yet been discovered. It was named "Achilles" after the Greek hero of the Trojan War. (Though asteroids are usually given feminine names, those with unusual orbits are given masculine names.)

Careful measurement showed that Achilles was moving at a speed of 8.1 miles per second, which meant it was as far from the sun as Jupiter was, for Jupiter moved at that speed too. Then Achilles was observed for a long enough period to calculate the shape of its orbit. Most asteroids had orbits that were more elliptical than those of ordinary planets. Achilles, however, was different. Its orbit was nearly circular.

But if Achilles is as distant as Jupiter and has an orbit as circular as Jupiter's, then Achilles had to be moving in Jupiter's orbit. And so it was! It was about 60 degrees ahead of Jupiter. That meant that the distance between Jupiter and Achilles was about 480 million miles, which is equal to the distance of each from the sun.

Jupiter, Achilles, and the sun are at the corners of an equilateral triangle. Back in 1772, Lagrange, working from gravitational theory, had calculated that such a situation was stable if the third body were very small compared to the other two. The three ought to form an equilateral triangle that would keep its shape as the triangle revolved about the apex at which the sun was located.

Well, Achilles *was* very small compared to Jupiter and the sun; it was only a hundred miles or so across. Therefore it ought to stay in place, always 60 degrees ahead of Jupiter as both traveled around the sun.

To be sure, Achilles was pulled this way and that by perturbations of other bodies, so that it slipped a few million miles in one direction or the other. On the average, though, it remained 480 million miles ahead of the giant planet.

But we can draw an equilateral triangle in the other direction, too. There was another stable point in Jupiter's orbit 480 million miles *behind* the giant planet. Was there any asteroid in that position?

Before the year was over, an asteroid was located in that position too. It was asteroid number 617 (twenty-eight other asteroids had been discovered in the months between) and it was named "Patroclus" after Achilles's great friend in Homer's tale of the Trojan War.

Astronomers realized at once that it was very likely that more than one asteroid might be in those positions. Any asteroid that blundered into a position 60 degrees before or behind Jupiter, being pulled or pushed in by perturbations and moving at about the right speed, would stay there from then on. It would be "captured" by Jupiter.

The search was on, and additional asteroids were found in each position. All were named after other characters in the *Iliad*.

In the position ahead of Jupiter, along with Achilles, are asteroids named Hector, Nestor, Agamemnon, Odysseus, Ajax, Diomedes, and Antilochus. Of these, all but Hector are named for Greek warriors. In the other position, along with Patroclus, are Priamus, Aeneas, Anchises, and Troilus. Of these, all but Patroclus are named for Trojans.

Because all these asteroids are named after characters in Homer's tale of the Trojan War, they are called the "Trojan asteroids." And because of this, any position which is at the third point of an equilateral triangle of which the other two points are occupied by large bodies is said to be a "Trojan position."

The sun, Jupiter, and the Trojan asteroids remain the only example of stable equilateral triangles in the sky. Might there be asteroids in the Trojan positions of any of the other planets? Perhaps; but if so, they are too distant or too small, or both, to be seen.

6 ●

THE
GRIP
OF
JUPITER

The Outer Satellites

There is still another and even *more* intimate way in which
Jupiter can be associated with the asteroids.

Consider that some of the asteroids are to be found as far out
from the sun as Jupiter's orbit. That is true of the Trojan
asteroids which ventured out that far and were trapped in the
Trojan position. (In fact, there is at least one asteroid, Hidalgo,
discovered in 1920, which recedes as far as Saturn's orbit.)

Could it be that some asteroids were drawn, through various
perturbations, into an orbit that brought them, very occasionally, within a few million miles of Jupiter itself? If so, it was
possible that Jupiter might actually capture the asteroid. Jupiter's gravitational grip might so alter the asteroid's orbit as to
cause it to circle Jupiter indefinitely.

Even before Wolf's discovery of Achilles showed that asteroids might exist as far from the sun as Jupiter was, an American astronomer, Charles Dillon Perrine, wondered if Jupiter
might have such captured asteroids as satellites. Such captured
asteroids were likely to be quite small and very dim. That would
be why they had not yet been discovered.

In 1904 Perrine began investigating the space near Jupiter
to see whether he could detect any very dim satellites. It was
not long before he discovered two, one in December 1904 and

the next in January 1905. They were the sixth and seventh satellites of Jupiter to be discovered.

These new satellites were quite different from those already known. For one thing, they were quite small. The sixth might be perhaps fifty miles across, the seventh, thirty-five. Then, too, they were much farther from Jupiter than were any of the others. They were over seven million miles from Jupiter, six times as far from Jupiter as was Callisto (the outermost of the Galilean satellites).

Perrine's discoveries encouraged other astronomers to search for small satellites of Jupiter. In 1908 P. J. Melotte, in England, found another satellite, the eighth.

The eighth satellite was even more unusual than the sixth and seventh. It was another tiny body, less than twenty miles across, most likely, and it was twice as far from Jupiter as Perrine's satellites were. Its average distance from Jupiter was well over fourteen million miles and it took two years to complete its orbit about the planet.

In later years four more tiny satellites of Jupiter were discovered—all by the American astronomer Seth B. Nicholson. He discovered a ninth satellite in 1914, a tenth and eleventh in 1938, and a twelfth in 1951.

Nicholson's satellites are all tiny bodies less than twenty miles across, and there is no reason to suppose they represent all the satellites there are. Jupiter's satellite number of twelve is almost certainly not complete; but the other satellites, assuming they exist, are too small and therefore too dim to be detected from Earth.

The outer satellites of Jupiter have no official names. These satellites, and Jupiter's innermost satellite (which I have been calling Amalthea) are the only satellites in the system—indeed the only known bodies of any sort—that lack names. Usually they are known by J and the Roman numeral that represents the order of discovery.

Names have been unofficially suggested, all of them of mythological beings associated with Jupiter (or Zeus).

Thus, J-VI is sometimes called Hestia, the name of a sister of Zeus, while J-VII is Hera, who is both Zeus's sister and his wife. J-VIII is Poseidon and J-IX is Hades, each a brother of Zeus. J-X is Demeter, a sister of Zeus; J-XI is Pan, a grandson of Zeus; and J-XII is Adrasteia, for a nymph who, like Amalthea, took care of Zeus in his infancy.

In Table 38 the average distances and the periods of revolution are given for each of the outer satellites, with their names given in parentheses as befits their unofficial status. Callisto, the outermost of the Galilean satellites, is included for comparison.

Notice that the outer satellites fall into two groups. There

TABLE 38

Distance and Period
of the Outer Satellites of Jupiter

SATELLITE		DISTANCE		PERIOD OF REVOLUTION	
		IN MILES	IN KILOMETERS	IN DAYS	IN YEARS
J-IV	Callisto	1,170,000	1,880,000	16.69	0.045
J-VI	(Hestia)	7,150,000	11,470,000	250	0.68
J-X	(Demeter)	7,300,000	11,710,000	263	0.72
J-VII	(Hera)	7,325,000	11,740,000	259	0.71
J-XII	(Adrasteia)	12,900,000	20,700,000	631	1.73
J-XI	(Pan)	13,900,000	22,350,000	692	1.90
J-VIII	(Poseidon)	14,500,000	23,300,000	738	2.02
J-IX	(Hades)	14,750,000	23,700,000	758	2.07

are three with an average distance from Jupiter of a little over seven million mlies, and four with an average distance of fourteen million miles or so.

This does not mean that the three and the four are in the same orbits at all. They would be, some of them, in nearly the same orbits if those orbits were circular and in the plane of Jupiter's equator, as is true of the inner satellites.

In the case of the outer satellites, however, the orbits are quite eccentric to begin with. This means that each orbit is a distinct ellipse, with its long axis pointed in some particular direction. Just because two or more satellites have the same *average* distance doesn't mean they have the same orbit at all.

In Table 39 the eccentricity of each of the outer satellites is given, together with their minimum and maximum distances from Jupiter.

J-VIII has the most eccentric orbit, and at its farthest point recedes fully twenty million miles from Jupiter. That is the farthest distance from the primary that any known satellite ever reaches. This means that when Jupiter and its satellites are at their closest to the Earth, we can make out J-VIII over 3° from Jupiter, a distance equal to six times the apparent width of our moon.

At the opposite end of its orbit, J-VIII comes within nine million miles of Jupiter and is almost as close to it as the J-VI, J-X, J-VII group.

If all the orbits are drawn on a piece of paper, there will be many points where one orbit crosses another. Does this mean that there is a chance of collision among the various outer satellites? Might each of two satellites be approaching the crossing point of their orbits at the same time?

No. That is not possible. The orbits only appear to cross because they are drawn on a piece of paper. An actual model of the orbits would have to be made in three-dimensional space —elliptical wires around a ball. Then it could be seen that

TABLE 39

Eccentricity of Jupiter's Outer Satellites

SATELLITE	ECCEN-TRICITY	MINIMUM DISTANCE		MAXIMUM DISTANCE	
		MILLIONS OF MILES	MILLIONS OF KILO-METERS	MILLIONS OF MILES	MILLIONS OF KILO-METERS
J-VI (Hestia)	0.155	6.1	8.3	9.7	13.2
J-X (Demeter)	0.08	6.7	7.9	10.8	12.6
J-VIII (Hera)	0.207	5.8	8.9	9.3	14.2
J-XII (Adrasteia)	0.155	10.8	15.0	17.4	24.1
J-XI (Pan)	0.21	11.0	16.8	17.7	27.0
J-VIII (Poseidon)	0.378	9.0	14.4	20.0	32.2
J-IX (Hades)	0.27	10.7	18.8	17.2	30.2

where two orbits seem to cross as drawn on a piece of paper, one is actually far above the other. In space, a million miles or more might separate satellites at what seems to be a crossing point.

In Table 40 the orbital inclination of each of the satellites is given. This is the amount by which the plane of its orbit is tilted to the plane of Jupiter's orbit. Jupiter's equator is tilted three degrees to its orbit, and so are the orbital planes of the Galilean satellites. Not so the planes of the orbits of the outer satellites. They revolve in planes that are strongly tipped.

Even when two or more satellites have inclinations of about the same amount, the tilting can nevertheless be in different directions. If two orbits are tilted through the same angle but in different directions, the high point of one is nowhere near the high point of the other.

The orbital inclinations of the four outermost satellites is particularly high. That raises an interesting point that can be best explained by considering a planet's rotation.

Suppose we are looking at a planet through a telescope. If its north pole is straight up and its south pole is straight down, then we might see it rotating from west to east. Jupiter rotates in this fashion. So does the Earth. It is the normal fashion of rotation.

But what if the planet's axis (as we see it through the telescope) is tilted a bit? Then the rotation still seems to us to be mainly west to east, but there is also some movement north to south.

The larger the axial tilt, the more the rotation is seen as going from north to south, and the less from west to east. When the axial tilt is 90°, the axis is horizontal, as we see it, and the planet is rotating from top to bottom—that is, from north to south.

If the axial tilt is a little greater than 90°, the rotation is

TABLE 40

Orbital Inclination of Jupiter's Outer Satellites

SATELLITE		ORBITAL INCLINATION (DEGREES)
J-VI	(Hestia)	29
J-X	(Demeter)	28
J-VII	(Hera)	28
J-XII	(Adrasteia)	147
J-XI	(Pan)	163
J-VIII	(Poseidon)	148
J-IX	(Hades)	157

still mostly north to south, but now it is also a little east to west.

In other words, if the axial tilt is less than 90°, the rotation is, at least in part, from west to east. If the axial tilt is more than 90°, the rotation is, at least in part, from east to west. Axial tilts are almost always less than 90°, so that we think of the west-to-east rotation as normal or "direct." The east-to-west rotation is "retrograde," from a Latin word meaning "backward." (Sometimes direct rotation is called "prograde" from a Latin word meaning "forward.")

If you look at Table 20 on page 66, you will see that the axial tilts of Uranus and Venus are greater than 90°. Both these planets rotate in retrograde fashion. All the rest, including Earth and Jupiter, rotate in prograde fashion.

All this also applies to the orbital inclination of satellites. In almost all cases, the inclination of satellites is less than 90°. This is true of our moon, for instance, and of the Galilean satellites of Jupiter. All revolve in prograde fashion. So do Amalthea and the three satellites of the inner group of Jupiter's small satellites.

The four outermost satellites of Jupiter, however, all have orbital inclinations of more than 90°, and all revolve about Jupiter in retrograde fashion.

There is no mystery about this. Astronomers are quite sure that the small outer satellites of Jupiter are captured asteroids. The four outermost satellites are just barely within Jupiter's grip and must have been captured only with difficulty.

When an asteroid approaches Jupiter (or any planet) at a distance that is just about the edge of its influence and where the planet can just barely overcome the pull of the sun, that asteroid is more easily captured if it can move into a retrograde orbit rather than a prograde one.

The three satellites forming the closer group of the outer satellites approached closely enough to be captured in prograde orbits. The four forming the more distant group of the outer satellites are so far from Jupiter that they could not have been captured in prograde orbit. It was just the chance that led them to approach Jupiter from a direction that made a retrograde orbit possible that led to their capture.

If the outer satellites were captured and have not been there since Jupiter's beginnings, it may be that they won't remain part of the system forever.

The sun's attraction is sufficiently powerful to introduce complex changes in their orbits. On occasion, perturbations from other planets—Saturn particularly—may just shift the orbit of J-VIII or J-IX in such a way as to make the sun's pull powerful enough to cause it to drift farther from Jupiter and still farther. Then the time may come when one of those satellites might take on, once again, an independent asteroid orbit.

And on the other hand, by that time Jupiter may have captured other asteroids and formed new satellites.

The Comets
We have now considered the sun, the planets, and the asteroids and seen how they are influenced by Jupiter's gravitational pull. Is there anything else?

Yes, there are the comets.

Some astronomers think that far out from the sun, some trillions of miles distant, hundreds of times as far away as Pluto, there is a shell of small asteroids, perhaps as many as a hundred billion in number. Each is a mile across or less, so that the total mass may not be much more than that of the moon.

Slowly, these distant asteroids circle the sun. Even at that distance, the sun's gravitational field is in control, but so feebly that it would take these asteroids 30 million years or so to complete one revolution.

These distant asteroids suffer perturbations from the gravitational effect of the nearer stars. Some might be pulled forward in such a way as to be made to take up an orbit still farther from the sun. Some might even be pulled free of the sun's influence altogether in this way. Others might be pulled back and made to take up an orbit closer to the sun.

Occasionally the effect of perturbations may pile up on a particular asteroid to the point where it is made to take up an orbit that directs it sharply toward the sun. As it moves toward the sun, it gains speed steadily. By the time it approaches the sun, it is streaking along at dozens of miles a second. It moves quickly around the sun and then goes zooming back out to its position far beyond Pluto.

It has a new orbit, a very elliptical one, and once this new orbit is formed, the asteroid stays in it until additional perturbations introduce new changes.

These very distant asteroids are made up not only of rocky substances as ordinary asteroids are, but in addition contain substances that would melt and vaporize at ordinary Earth-like temperatures. Ordinary asteroids are so close to the sun that if they had had such substances as part of their structure, the sun's warmth would long since have melted and vaporized them away. Far out in space, distant from the sun, these substances remain frozen forever, however, and can make up part of the structure of the distant asteroids.

As a distant asteroid approaches the sun, however, and as its temperature goes up, these substances vaporize. The asteroid begins to be enveloped by a cloud of gas. In addition, the frozen substances had acted as cement holding rocky dust and pebbles together. As the substances evaporate, the dust and pebbles come loose too.

The sun is always sending out streams of high-speed particles in all directions, and these make up the "solar wind." The solar wind strikes the gas and dust around the incoming asteroid and forces it out, away from the sun. The asteroid,

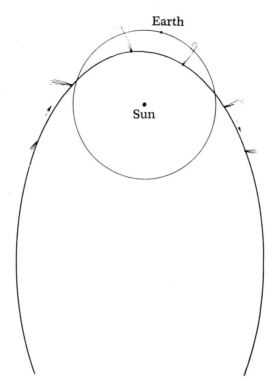

A comet's tail, consisting of dust and vapors pushed away from the sun by the solar wind, trails behind the comet as it approaches the sun, but whips around and precedes the comet as it recedes from the sun.

whose feeble gravity can't hold the gas and dust very well, trails a long tail of it in the direction away from the sun. The material in that tail is lost forever to the asteroid.

The asteroid, once it comes close enough to Earth to be seen, appears as a faintly shining, fuzzy object trailing a long tail. The tail seemed to resemble streaming hair, and the object was called a "comet" from a Latin word for "hair."

Once a comet is forced into an orbit that sends it close to the sun, it doesn't last long. Each time it approaches the sun it loses dust and gas, and after a few hundred approaches nothing is left but a small rocky core at most. The dust and gas then spread out to fill the orbit of the comet. It is this rocky dust, probably, that is responsible for the many millions of meteoric fragments (a pinhead in size or less) that enter the Earth's atmosphere every day.

A comet that drops toward the sun from the far-distant shell may take millions of years to reach the sun, swing around, and return. Such a comet is a one-time sight as far as men are concerned. If such a comet were to flash across our skies this year, no one could expect to see it again in tens of thousands of lifetimes.

Such comets are usually quite bright and spectacular because they have only passed through the inner solar system a few times and still have much material with which to form the cloud of vapor and dust that becomes the long, bright tail.

Sooner or later, though, as a comet passes through the inner solar system, it will pass close enough to a planet to be strongly perturbed. The planet's gravitational field causes the comet to curve slightly around the planet in its path, and after that its orbit is changed.

It frequently happens that the effect of the planetary perturbation is to curve the comet's path in such a way as to shorten its orbit. It no longer will recede so far from the sun. In fact, its entire orbit may remain within the planetary portion of the solar system, and it may return to the vicinity of the sun every hundred years or less. It has been "captured" and has become a "short-period comet."

The first astronomer to study the orbit of a comet in detail was an Englishman, Edmund Halley. He noticed that a number of comets recorded in past history had crossed the same part of the sky, and that they were separated in time by about 75 years. He decided that it was the same comet following an orbit around the sun and returning again and again.

In 1705 he announced this theory and predicted that the comet would return in 1758. It did, being first sighted on Chirstmas Day of that year. It has been called "Halley's comet" ever since. It returned again in 1835 and 1910, and will be seen on its next return in 1986.

The orbit of Halley's comet is enormously elliptical by planetary standards. At its closest approach to the sun, Halley's comet is only 54 million miles from it. It is then closer to the sun than Venus is. At the opposite end of its orbit, however, it is 3,200 million miles from the sun, farther out than the planet Neptune. Its eccentricity is 0.967.

Although Halley's comet moves from within Venus's orbit to beyond Neptune's, it doesn't cross any of those orbits. Its orbit is inclined to the plane of Earth's orbit by 28 degrees, so that it never passes closer than many millions of miles to any of the planets.

Even so, comets are affected by the pull of even a distant planet. For instance, Halley's comet, returning in 1758, was 69 days late and didn't reach the sun's vicinity till sometime in 1759. This was because the distant drag of Jupiter had held it up. If Jupiter had been the only planet involved, the delay would have been 135 days. However, the gravitational pull of Saturn, Uranus, and even Earth had worked in the opposite direction and had canceled 66 of those days.

Halley's comet is the most famous of all the short-period comets. No comet has appeared in 20th-century skies brighter than Halley's comet at its last appearance in 1910. This is the way its head looked at that experience, as seen through a good telescope.

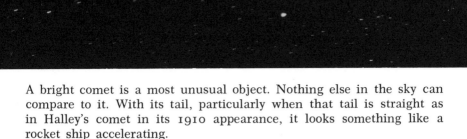

A bright comet is a most unusual object. Nothing else in the sky can compare to it. With its tail, particularly when that tail is straight as in Halley's comet in its 1910 appearance, it looks something like a rocket ship accelerating.

If a comet makes a particularly close approach to a planet, it may make such a sharp swerve that in the future its orbit will always carry it back to the point where that swerve took place. In that case, the farthest point in its orbit will be close to the orbit of the planet that affected it.

Thus Halley's comet, with its farthest point somewhat beyond Neptune's orbit, might have been swerved into its present orbit by Neptune. Sometimes it is spoken of as a member of the "Neptune family" for that reason. However, because of the orbital inclination of Halley's comet, it is more than a billion miles from Neptune's orbit when both bodies are equally far from the sun, and it seems doubtful that it could have been affected by Neptune sufficiently to be captured.

Halley's comet is much closer to Jupiter's orbit when both bodies are equally far from the sun. Halley's comet can approach as close as 200 million miles to Jupiter, and Jupiter is, of course, much more massive than Neptune and has a much stronger gravitational field. Perhaps it was Jupiter's influence that shortened the orbit of Halley's comet.

In fact, some astronomers think that only Jupiter is massive enough to have much chance of producing really major effects on the orbits of comets. They think that every comet that enters the planetary portion of the solar system will sooner or later be captured through Jupiter's influence.

There is certainly a "Jupiter family" of comets. More than fifty comets are known to have had their orbits seriously disturbed by Jupiter. Portions of the orbits of these comets are close to the orbit of Jupiter—a sure sign.

Perhaps the most unusual member of the Jupiter family of comets is "Encke's comet." It was discovered in 1818 by the French astronomer Jean Louis Pons, but the orbit was calculated the next year by the German astronomer Johann Franz Encke, and it is the latter whose name is used for the comet.

Encke's comet has the shortest period of revolution of any known comet. It completes its circle about the sun in 3.3 years. It approaches within 31 million miles of the sun at perihelion. This is quite close to Mercury's orbit and, indeed, the perturbing effect of Mercury on Encke's comet has been used to calculate the mass of that small planet.

At aphelion, Encke's comet is 380 million miles from the sun, so that it never quite recedes as far as Jupiter's orbit. No other comet has an aphelion so close to the sun.

Encke's comet has been observed at every revolution since its discovery, so that it has been seen to make some fifty approaches to the sun. Naturally, it is small and faint. Most of the vapors and dust it may have had to begin with have long since been driven away by the solar wind. What dregs remain just barely serve to make it visible.

In fact, all the members of Jupiter's family of comets are small and faint. All have periods of nine years or less, so that all have made many approaches to the sun and all have lost the dust and gas needed to make a comet spectacular.

We don't have to theorize that Jupiter can affect a comet's

orbit in a major way. We can actually see it happen sometimes.

The Pons-Winnecke comet had an orbit that was actually enlarged while astronomers were watching. During the nineteenth century the perihelion distance of the Pons-Winnecke comet increased steadily. In 1815 the comet approached as closely as 72 million miles to the sun, nearly as closely as Venus does. Now, the perihelion distance is 105 millions miles. It no longer approaches the sun even as closely as Earth does.

In 1889 the Brooks comet was discovered, with a perihelion distance of 175 million miles (somewhat outside the orbit of Mars) and an aphelion distance at 500 million miles (quite close to Jupiter's orbit). Its period was just about seven years.

Why wasn't it seen earlier?

Calculations showed that this was a brand-new orbit. Before its discovery, it had passed within 55,000 miles of Jupiter and had swerved sharply. Before then, its period had been 29 years and it was too far from Earth at all times to be noticeable.

Encke's comet is a dim one. It is fuzzy because some dust and vapor are heated off its surface, but not enough is produced to form a tail. The stars are seen here as rather long tracks because the photograph must be exposed for quite a while to record the comet, and in that time the camera must follow the comet and not the turning stars.

Even more spectacular is the case of Lexell's comet, named for a Swedish astronomer, Anders J. Lexell, who discovered it in 1770. At that time it was traveling in an orbit with a period of 5.5 years, but that, too, was a brand-new orbit. In 1767 it had passed through Jupiter's satellite system and had been swerved into the new orbit.

In passing through the satellite system of Jupiter, it did not affect the motions of Jupiter or the Galilean satellites in any detectable way. Why should it? It was just a tiny hunk of matter about a mile across or so. In fact, Lexell's comet was slowed up slightly when it passed Earth, so that it lost 2½ days in completing its orbit. Earth wasn't affected either.

When Lexell's comet approached its aphelion point again after making two more revolutions, it had the ill luck to pass fairly close to Jupiter again. Once more the comet swerved, and this time took up an orbit of a much longer period. It has not been spotted since; or at least if it has, it has been considered a new comet, there being nothing to connect it with the old comet of 1770.

Surface Gravity

Let's turn back to the planets and satellites themselves.

If we wish to know the gravitational effects of bodies at a considerable distance, all we have to do is to compare their masses. If one body has a hundred times the mass of another, the first will exert a hundred times the gravitational effect of the other at equal distances.

As an example, the Earth has 81 times the mass of the moon. Therefore a body which is a hundred thousand miles from the Earth's center is held by the Earth with 81 times the force that it would be held by the moon if it were a hundred thousand miles from the moon's center.

The situation is a little more complicated, though, if we

consider what the gravitational force is at the *surface* of the Earth, as compared with what it is at the *surface* of the moon. (Now we are speaking of "surface gravity.")

The Earth may be exerting 81 times the gravitational effect of the moon *at equal distances*, but the surfaces of the two bodies are not at equal distances from the centers. The Earth's surface is 3,950 miles from its center; the moon's surface is only 1,080 miles from its center. As the distance from the center decreases, the gravitational effect grows stronger in proportion to the square of the distance, and vice versa.

The surface of the Earth is 3.65 times as far from its center as the surface of the moon is. There is a weakening effect to Earth's gravitational pull, then, that is equal to 3.65 × 3.65, or 13.4. In comparing the surface gravity of Earth and moon, therefore, we must divide the mass ratio of 81 by 13.4, which leaves a ratio of 6 in Earth's favor.

In short, the gravitational pull on the moon's surface is ⅙ that of the Earth's pull on *its* surface. If we are talking about surface gravity, we can say that the moon's gravitational pull is ⅙ that of the Earth.

Calculating from the mass and size of each planet, we can work out the surface gravity in each case, as given in Table 41. (We are supposing, in Table 41, that the surface of each planet, as we see it in the telescope, is as solid as Earth's and that we could stand on it if we chose. This is certainly not true of the Jovian planets, but we'll ignore that for now.)

It is not surprising that Jupiter has the largest surface gravity of any planet. It is perhaps a little surprising that the surface gravity isn't greater than it is, considering that Jupiter is 318 times as massive as Earth. It may be even more surprising that the other Jovian planets have surface gravities scarcely larger than that of Earth. The surface gravity of Uranus is just about equal to that of Earth.

TABLE 41

Surface Gravity of the Planets

PLANET	SURFACE GRAVITY (EARTH = 1)
Mercury	0.38
Venus	0.90
Earth	1.00
Mars	0.38
Jupiter	**2.68**
Saturn	1.15
Uranus	0.99
Neptune	1.28
Pluto	0.4

In every case, of course, the visible surface of the Jovian planets is considerably farther from the planetary center than Earth's surface is. That weakens the gravitational pull at the visible surface of Jupiter and the other Jovian planets. The size of the outer planets and, therefore, the distance from the surface to center, is even larger than it might otherwise be, because the low density of those planets makes it large for its mass.

Even so, a surface gravity of 2.68 is serious enough. Suppose we imagine astronauts standing on Jupiter's visible surface. If one of them weighed 150 pounds on Earth, he would weigh 400 pounds on Jupiter. What's more, every 15 pounds of equipment he had to carry, such as his spacesuit and oxygen pack, would weigh 40 pounds on Jupiter.

That sounds bad enough, but there's worse to come—

Let's consider the Earth once again. Suppose you throw a ball up in the air. It rises to a certain height and then it comes down again. If you throw it harder, it will go higher before coming down again.

You might think that no matter how hard you throw the ball, it will always eventually come down, but that is wrong. The point to remember is that Earth's gravitational pull weakens with distance. If you throw a ball hard enough, it will rise so high that Earth's gravitational pull will have weakened considerably and then the ball will continue to rise higher than you might expect.

In fact, if you threw the ball fast enough, it might rise so high that the gravitational pull of the Earth would weaken more rapidly than the ball's speed would decrease. In that case, the ball would never come down again.

At what speed must you throw the ball to begin with, in order to make it rise in such a way that gravity weakens too quickly to bring it back? On Earth it turns out that the initial speed must be at least 7.0 miles per second. Anything moving at that speed or more leaves the Earth and never comes back. We speak of 7.0 miles per second as the "escape velocity" of Earth.

(A human arm can't throw anything with that speed. That is why everything we throw does come back to Earth. Even a gun or a cannon can't fire anything at that speed. Only a rocket can build up such speeds, little by little. That is one reason we use rockets to take men to the moon.)

On a planet like Jupiter, the escape velocity must naturally be higher than on the Earth. To begin with, the surface gravity is 2.68 times that of Earth.

Worse than that, not only is Jupiter's gravitational pull higher at the surface than the Earth's is, but Jupiter's pull weakens more slowly with distance.

The gravitational pull of any body weakens in proportion to the square of the distance from its center. At Earth's surface, the ball is 3,950 miles from Earth's center. If it travels 3,950 miles up from the surface, it has doubled its distance from Earth's center, and Earth's gravitational pull on it is only ¼ as strong as at Earth's surface.

On Jupiter, however, the surface (at the equator) is 44,000 miles from Jupiter's center. If a ball travels 3,950 miles up from the surface, its distance from the center has only increased by less than 10 percent, and the gravitational pull of Jupiter is still 90 percent of what it was on the surface. A ball must travel 44,000 miles up from Jupiter's surface before the strength of Jupiter's gravitational pull falls to ¼ what it was at its surface.

The escape velocity of Jupiter is higher than the Earth's, then, not only because of Jupiter's large surface gravity, but because of the slowness with which that surface gravity decreases in strength. The result is that the escape velocity from Jupiter is over five times that of the escape velocity from Earth. In Table 42, the escape velocities from all the planets are given.

The difficulty of a Jupiter landing, then, depends not only on the weight of the astronauts on Jupiter. It depends also on the enormous quantities of fuel that would be necessary to brake the speed of the spaceship and keep it from falling too fast under the pull of Jupiter's gravity. It would depend also on the need for enormous quantities of fuel to build up a speed over five times that required on Earth in order to pull away from the giant planet.

It's certainly risky to say that because of this man can *never* land on Jupiter (even though Jupiter's gravitational pull is only one of the problems and we'll come to others later on), but certainly we're not anywhere near ready to do it now.

So far, we have managed to travel to the moon, land on it, and return, but remember that Jupiter is 1500 times as far away as the moon is, even when Jupiter makes its closest approach. What's more, Jupiter has 17 times the surface gravity the moon has, and 25 times the escape velocity.

Steppingstones

But we don't have to be able to land on Jupiter to be able to gain close-up information about it.

For one thing, we can send an unmanned rocketship out toward Jupiter (a "Jupiter probe"). It can carry instruments that will record various kinds of measurements and radio them back to Earth. Such a probe was indeed launched in 1972.

We might even establish manned bases *near* Jupiter, rather than on it, once we learn how to support men on the long trip to Jupiter and back. After all, men don't have to land on Jupiter itself to be able to study it. There are steppingstones on the way to Jupiter in the form of Jupiter's satellites.

Do we need to be much concerned with the gravitational pulls of the satellites themselves? The small ones have only tiny gravitational pulls, and even the Galilean satellites, which

TABLE 42

Escape Velocity From the Planets

	ESCAPE VELOCITY		
PLANET	MILES PER SECOND	KILOMETERS PER SECOND	EARTH = 1
Mercury	2.6	4.2	0.37
Venus	6.4	10.3	0.91
Earth	7.0	11.3	1.00
Mars	3.1	5.0	0.44
Jupiter	**37.6**	**60.5**	**5.35**
Saturn	22.0	35.2	3.1
Uranus	13.5	21.7	1.9
Neptune	15	24	2.1
Pluto	3	5	0.4

are moon-size, must have surface gravities similar to that of the moon, and that can surely be handled. Table 43 gives the surface gravity of the Galilean satellites as compared with the moon.

And what about the escape velocities? Are there any unpleasant surprises there? In Table 44, we have the escape velocities of the Galilean satellites as compared with the moon. As you see, both in surface gravity and in escape velocity the Galilean satellites are very moonlike.

Very likely their surfaces are like that of the moon in some respects, too. It is hard to see any details of the surfaces of the Galilean satellites because of their huge distance from us and because of the glare of nearby Jupiter. Astronomers are pretty sure, however, that they have no atmosphere, any more than our moon does. There are also some indications that the surface is generally rough, like that of the moon, but in December 1972, astronomers at M.I.T. reported evidence that the Galilean satellites, *unlike* the moon, are covered with considerable frost.

In other words, if any one of the Galilean satellites were substituted for our moon, we would have no particular trouble in landing and taking off from it once again.

The situation would be even simpler in the case of the small

TABLE 43

Surface Gravity of the Galilean Satellites

SATELLITE		SURFACE GRAVITY (MOON = 1)
J-I	Io	1.1
J-II	Europa	0.9
J-III	Ganymede	1.0
J-IV	Callisto	0.7

TABLE 44

Escape Velocity From the Galilean Satellites

	ESCAPE VELOCITY		
SATELLITE	MILES PER SECOND	KILOMETERS PER SECOND	MOON = 1
J-I Io	1.5	2.4	1.0
J-II Europa	1.3	2.1	0.9
J-III Ganymede	1.8	2.9	1.2
J-IV Callisto	0.9	1.4	0.6

asteroids, if these were considered all by themselves.

Amalthea (J-V) is, for instance, about 70 miles across. If we assume it is made up of material like that which makes up our own moon, its surface gravity would be only about $\frac{1}{30}$ that of the moon or about $\frac{1}{180}$ that of the Earth. A 150-pound astronaut would weigh only 13 ounces on Amalthea. On the smallest of the outer satellites of Jupiter, he would weigh less still, perhaps only about an ounce or two. Escape velocities would also be very small.

The trouble is, though, that none of these satellites are alone in space. There is giant Jupiter in the vicinity. What about *its* gravitational pull? Is it possible to build up a rocket velocity that would get a spaceship off Ganymede, for instance, and yet leave it still trapped by the gravitational pull of Jupiter? Would it be free of Ganymede only to find itself in orbit about Jupiter?

To answer that question, it is possible to calculate the escape velocity from Jupiter at the distance of each of the satellites, and this is given in Table 45.

For the small satellites, with tiny escape velocities of their own, it is the pull of Jupiter that must be taken into account.

If you compare Tables 44 and 45, you will see that even in the case of the Galilean satellites, Jupiter's pull is far more important than that of the satellites themselves.

From Amalthea, a velocity of nearly 24 miles a second must be achieved to get a spaceship away from Jupiter. At any lesser speed, the ship will get away from Amalthea (not at all difficult) but remain in an orbit about Jupiter, a satellite itself. Remember, too, that 24 miles per second is over three times the

TABLE 45

Escape Velocity From Jupiter at Satellite Distances

	ESCAPE VELOCITY FROM JUPITER	
AT DISTANCE OF	MILES PER SECOND	KILOMETERS PER SECOND
J-V Amalthea	23.8	38.2
J-I Io	15.4	24.8
J-II Europa	11.9	19.1
J-III Ganymede	9.8	15.7
J-IV Callisto	7.0	11.2
J-VI, J-X, J-VII (average)	2.8	3.1
J-VIII, J-IX, J-XI, J-XII (average)	2.1	2.4

velocity required to escape from Earth. Even at the distance of Callisto, the escape velocity from Jupiter is equal to that from Earth.

This is not as bad as it sounds, however. Advantage can be taken of the speed of the satellites themselves. Amalthea moves in its orbit at a speed of 17.2 miles per second. If the spaceship takes off in the direction of that motion, it has a speed of 17.2 miles per second to begin with. If it then adds a speed of its own of 7 miles per second (the kind of speed it needed to get free of the Earth), it will have reached a speed more than enough to escape from Jupiter.

Ganymede moves at 6.8 miles per second. The spaceship would have to add only 3 miles per second of its own, if launched in the direction of Ganymede's motion, to pull away not only from Ganymede but from Jupiter as well.

Still, there is no question but that maneuvering as close to Jupiter as the Galilean satellites find themselves will be a tricky thing. It may well be better to nose cautiously toward the small outer satellites. Consider J-VIII, which has only a tiny escape velocity of its own. At its greatest distance from Jupiter (20 million miles), a spaceship would require a speed of a little less than 1.8 miles per second to escape from Jupiter, only about a quarter of the speed required to escape from Earth.

It may well be, then, that the first manned spaceflights to Jupiter will choose to play it safe and to conserve fuel by going no closer to the giant planet than its outermost satellites, the small captured asteroids.

Perhaps as we build up bases on those small bodies and develop more energetic motors and more sophisticated methods of driving rocketships, we will be able to penetrate closer and closer to the planet—moving down the line of the stepping-stone satellites.

7

THE
SIGHTS
OF
JUPITER

The Sun in the Satellite Sky

What will we see from the vantage point of the satellites, once we are there?

For one thing, we will see the sun in much shrunken form. Jupiter and its satellites are over five times as far from the sun as we are, and the sun would have a diameter of only about 6 seconds of arc instead of its diameter of 32 seconds as seen from Earth. The sun would cast only about $\frac{1}{25}$ as much light and heat on Jupiter and its satellites as it does on Earth and its moon.

The tiny sun would be made out as a small circle of light. It would be important to remember, though, that the lessening of its light would be entirely due to the smaller area it took up in the sky. The small circle would be just as bright as an equal area of our own sun in Earth's sky, and it could damage the eyes, small as it is, if looked at directly.

(Even if the sun were so distant that it could be seen only as a dim and tiny star, its brightness would be just as great as an equally large speck of the sun as we see it from Earth. In that case, though, the speck would be so tiny that the total light entering the eye would not be enough to do damage.)

The sun would seem to move across the sky as the satellites

rotated on their axes, just as it seems to move across our own sky. In order to know the details of its motion, however, we would have to know something about the rate of satellite rotation.

As far as Jupiter's small outer satellites are concerned, we have no information on their rate of rotation, so we can say nothing about the apparent motion of the sun in their skies.

In the case of Amalthea and the Galilean satellites, however, it seems pretty safe to assume that they always face one side toward Jupiter. Gravitational effects of a large body on a smaller one tend to make the smaller body turn in such a way as to face the same face to the larger body as it revolves about the larger body. Earth's gravitational pull has done this to our moon, which presents us always with the same face. Jupiter's much stronger gravitational pull would certainly have done this to its inner satellites.

When a satellite always presents the same face to the planet it circles, then its period of rotation must be exactly equal to its period of revolution.

Callisto, for instance, which revolves about Jupiter in 16.7 days, must rotate on its axis in 16.7 days if it always presents the same face to Jupiter. This means that if you were standing on the surface of Callisto, you would see the sun make a complete circle of the sky in 16.7 days.

It would rise in the east because Callisto revolves in prograde fashion. If Callisto revolved in retrograde fashion, it would rise in the west. The sun rises in the east as seen from all the inner satellites.

It would take the sun, as seen from Callisto, a little over four Earth days to rise to its highest point in the sky, and then a little over four more Earth days to sink to sunset. Each day and each night would be eight and a third Earth days long.

Those satellites which are closer to Jupiter than Callisto is revolve about it more quickly, and therefore rotate on their axes more quickly. The lapse of time from sunrise to sunset on each of the inner satellites is given in Table 46. The Earth and moon are included in the table for comparison.

During the night (provided Jupiter is not in the sky), the stars look very much as they do when seen from Earth. All the constellations are there in their familiar patterns, with the bright stars marking out Orion and the Big Dipper and all the rest. (The stars are so enormously far away that the distances between the planets are tiny in comparison. Therefore the pattern of stars doesn't change noticeably from one end of the solar system to the other.)

Since there is no atmosphere on the satellites, there are never any mists or clouds to hide the stars. They don't twinkle, and each star is a little brighter than it ever is when seen from Earth, since even the clearest air absorbs some starlight.

This means that some stars just too faint to be seen from Earth would be just visible on the Jovian satellites. The total number of stars we can see from Earth's surface with the un-aided eye is about six thousand. From Jupiter's satellites we might double the number, making out six thousand very dim stars we could not see on Earth.

TABLE 46

Length of Day and Night on the Inner Satellites

| | TIME BETWEEN SUNRISE AND SUNSET OR BETWEEN SUNSET AND SUNRISE | |
SATELLITE	IN HOURS	IN DAYS
J-V Amalthea	6.0	0.25
J-I Io	21.2	0.9
J-II Europa	42.7	1.8
J-III Ganymede	84.0	3.5
J-IV Callisto	200.2	8.3
Earth	12.0	0.5
Moon	354	14.8

The pattern of stars will vary from night to night. As Jupiter and its satellites move around the sun, the sun will be seen in a slightly different portion of the sky each day, and a slightly different portion of the sky will be visible each night. This happens on Earth too. Since the Earth moves about the sun in one year, we see the pattern come full cycle in one year. Jupiter takes nearly twelve years to circle the sun, so it takes nearly twelve years for the pattern of stars as seen by night to come full circle.

The Jovian satellites are without atmosphere, so the sunrise does not turn the sky blue. There is no air on the satellites to scatter blue light. The sky on the satellites remains black even with the sun in the sky, and some of the stars, at least, remain visible. The glare of the sun will probably make it difficult to see the dim ones, however.

Jupiter in the Satellite Sky

So far it doesn't seem that the sights in the sky, as seen from the satellites, are very astonishing. There is a shrunken sun, far less impressive than Earth's sun, and numerous new, but very dim, stars. The day and night would be more prolonged (except on Amalthea) than they are on Earth, but that's not much.

In fact, everything I have described so far can be seen, in very similar fashion, from our own moon.

But what about Jupiter? How would *it* look in the sky of the satellites?

Since Callisto (and the other inner satellites) always presents the same side to Jupiter, Jupiter does not seem to move in Callisto's sky (or in that of the other inner satellites). If you were standing on Callisto's surface, somewhere on the side facing Jupiter, you would see Jupiter motionless in the sky. It would neither rise nor set, but would always be present in just about the same place.

On the other hand, if you were standing on Callisto on the side away from Jupiter, then Jupiter would *never* be in the sky.

In fact, if you could imagine an intelligent creature on Callisto who was rooted to the ground like a tree and who lived on the side away from Jupiter, it might never know that Jupiter existed. Of course, it might work out the existence of Jupiter by analyzing Callisto's motions and concluding that *something* had to be pulling at it.

If you traveled across the surface of Callisto, Jupiter would seem to move in the sky in the direction opposite to that in which you were moving. If you walked far enough, you might see Jupiter reach the horizon. If you then moved no more, you would be able to view a permanent "Jupiter-set." On the other hand, if you were on the side away from Jupiter and traveled in some particular direction, you would eventually see a "Jupiter-rise" at that part of the horizon toward which you were heading.

So far this is rather similar to something we could see from our own moon. From the surface of our moon, Earth would seem to be in just about the same spot in the sky forever.

Jupiter, as seen from Callisto, would be 260′ in diameter, however. This is over eight times the width of our full moon as seen from Earth, and over twice the width of the Earth as seen from the surface of the moon. Jupiter in Callisto's sky would be a *much* more impressive sight than the Earth in the moon's sky.

Naturally, Jupiter would seem smaller when viewed from the small outer satellites, and would appear still larger when viewed from those satellites which lie still closer to Jupiter than Callisto does. In Table 47, the apparent width of Jupiter is given as seen from each of its satellites.

Even at the distance of the small outermost satellites, Jupiter would appear roughly as the moon does to us. It would seem a little smaller than our moon when viewed from the four farthest satellites, and a little larger than our moon when viewed from the three small satellites that are closer to Jupiter.

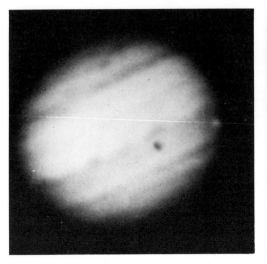

These two photographs were taken of Jupiter in infrared light. The one on the right shows the Great Red Spot. In both photographs a satellite is seen crossing the face of Jupiter. In the one on the left, the satellite casts a visible shadow (to its left), which is almost lost in the dark band.

From the Galilean satellites, however, it would dominate the sky as nothing dominates it here on Earth. From Amalthea, in particular, Jupiter would stretch across a distance that would take it from the horizon halfway to the zenith. It would fill up $\frac{1}{30}$ of the entire overhead sky.

And how bright would Jupiter appear as seen from the various satellites?

Since Jupiter is farther from the sun than the moon is, it shines less brilliantly for a given size of area exposed, even though Jupiter reflects six times as much of the light it receives as the moon does.

For instance, from J-VI, Jupiter would appear slightly larger than our moon does to us. Jupiter would have 1.3 times the diameter of our own moon and therefore an area 1.7 times our moon. Nevertheless, because Jupiter is bathed in the light of a far more distant sun, it ends up appearing only about $\frac{1}{3}$ as bright when seen from J-VI as our full moon does to us.

As we move down the line of satellites, and as Jupiter ap-

TABLE 47

Diameter of Jupiter as Seen From the Satellites

| | DIAMETER OF JUPITER | |
AS SEEN FROM DISTANCE OF	IN DEGREES	MOON AS SEEN FROM EARTH = I
J-VIII, J-IX, J-XI, J-XII (average)	0.4	0.7
J-VI, J-VII, J-X (average)	0.7	1.3
Callisto	4.3	8.3
Ganymede	7.3	14
Europa	11.7	25
Io	19	38
Amalthea	46	92

TABLE 48

Brightness of Jupiter as Seen From the Satellites

AS SEEN FROM DISTANCE OF	MAXIMUM BRIGHTNESS OF JUPITER (MAXIMUM BRIGHTNESS OF MOON AS SEEN FROM EARTH = I)
J-VIII, J-IX, J-XI, J-XII (average)	0.08
J-VI, J-VII, J-X (average)	0.3
Callisto	12.5
Ganymede	35
Europa	85
Io	220
Amalthea	1200

pears larger and larger, it would naturally get brighter and brighter. We see this in Table 48.

From the Galilean satellites, then, Jupiter appears considerably brighter, thanks to its huge apparent size, than our moon does to us. Even from Callisto, Jupiter shines with twelve and a half times the brilliance of our moon, and from Amalthea it is twelve hundred times more brilliant.

How does the great globe of Jupiter compare with the brightness of the shrunken sun as seen from the various satellites?

The comparison, you may be surprised to hear, is all in favor of the sun.

Even at the distance of Jupiter, the sun remains 17,000 times as bright as the full moon does to us. As a result, on every one of Jupiter's satellites that little pea-sized sun remains much brighter than Jupiter—as shown in Table 49.

Seen from Callisto, the sun remains well over a thousand times as bright as the large planet. Even on Amalthea, the huge swollen globe of Jupiter is never more than $\frac{1}{14}$ as bright, altogether, as the tiny sun.

The Phases of Jupiter

There's something important to remember in connection with Jupiter's brightness. Jupiter, like our moon, shines only by reflecting the light from the sun. Only half of its globe is lit by the sun and, depending on where the sun is in relation to Jupiter and the satellite, the part of Jupiter that faces the satellite may be entirely light, entirely dark, or partly light and partly dark.

In other words, Jupiter will show phases, when seen from its satellites, in just the same fashion that our moon does when it is seen from Earth (or as Earth does when it is seen from the moon).

When all the surface of Jupiter facing the satellite is lit by the sun, then Jupiter shines like a great round globe and is at

TABLE 49

Brightness of Sun Compared to Jupiter

SATELLITE	APPARENT BRIGHTNESS OF SUN (BRIGHTNESS OF JUPITER AS SEEN FROM THAT SATELLITE = 1)
J-VIII, J-IX, J-XI, J-XII	210,000
J-VI, J-VII, J-X	56,000
Callisto	1,350
Ganymede	490
Europa	200
Io	77
Amalthea	14

its maximum brightness. In Tables 48 and 49 the brightness of Jupiter is calculated at this maximum. Most of the time, however, only part of Jupiter's surface (as seen from the satellite) is bathed in sunlight. At those times Jupiter is dimmer. The less of its visible surface that is lighted, the dimmer it is, and there can be times when the surface facing the satellite receives no sunlight at all and Jupiter's brightness is just about zero.

Let's see how this works by imagining ourselves on Callisto again. We can imagine ourselves on a spot such that the sun is directly overhead, at the zenith. Naturally, it will stay there as long as we remain at that spot on Callisto's surface.

As we stand there, we will see the sun rise in the east, cross the sky, and set in the west 8⅓ days later. Then, another 8⅓ days later, it would rise again and repeat the process.

But while the sun is passing through the heavens, what happens as far as Jupiter is concerned?

Imagine it to be sunrise on Callisto. The tiny sun is shining at the eastern horizon and lights up Jupiter from the east. Only the eastern half of Jupiter, as we gaze at it in the sky, is lit. The western half gets no sunlight and is dark. Therefore we see a "half-Jupiter" (like our own half-moon, but, of course, considerably larger and brighter).

As the sun slowly rises in the sky, more and more of its light shines on the portion of Jupiter that is away from us and that we cannot see; less and less shines on the portion we can see. The lighted portion of Jupiter shrinks and becomes a crescent on the eastern side. The crescent is fat at first, but grows thinner as the sun rises higher and higher.

Finally, after four days, the sun is almost at zenith and is therefore very close to Jupiter. It isn't going to touch Jupiter, of course, for the sun is much farther away, but now the sunlight is shining almost entirely on the other side of Jupiter, the one we can't see. Only the thinnest possible crescent on the east can be seen, and the only way you can see it is to block out the light of the tiny sun, which would otherwise wash out the thin curve of Jupiter-light. The sun then passes behind Jupiter and is eclipsed.

It will take the sun about 4.5 hours to pass from one side of Jupiter to the other, as seen from Callisto. During all that time, all the light of the sun will be on the other side of Jupiter, and the side we will see will be completely dark. It will be the phase of "new-Jupiter."

Yet Jupiter won't be actually *all* dark. It has an atmosphere, and the atmosphere all around the rim of the planet will glow with sunlight passing through it. There will be a ruddy circle of light marking the rim of Jupiter. Within that circle, the darkness will be different from the darkness of the rest of the sky, for there will be no stars visible within the circle.

At the beginning of the eclipse, the eastern half of the circle of light will be lighter than the western half. In the middle of the eclipse, the circle will be evenly bright all around. Then the western half will gradually brighten while the eastern half will dim.

Finally, at the end of the eclipse, the sun will appear like a bright diamond at the western side of Jupiter. Slowly it will retreat, moving farther west, and there will be a crescent Jupiter again, on the western side this time, growing fatter and fatter. By the time the sun has reached the western horizon, nearly a hundred hours after the eclipse, the planet will be in the "half-Jupiter" phase again, but now it will be the western half of Jupiter that will be lit up.

Then the sun sets. It will no longer be shining on the part of Callisto where you are standing, but it is still shining on Jupiter. In fact, it is shining more and more on the side of Jupiter that is facing Callisto. The half-Jupiter begins to expand and approaches more and more closely to the full. Finally, a hundred hours after sunset, it is midnight on Callisto at the spot where you are standing. The sun, Callisto, and Jupiter are then in the same straight line. The sun shines past Callisto on all sides and lights up the complete side of Jupiter that is facing you. Jupiter is now a huge round circle of light; it is "full-Jupiter."

But then the sun passes the midnight point, and the western side of Jupiter begins to grow dark. The round circle of light begins pulling in from the western side, and by the time the sun is about to rise again, Jupiter is once again a half-Jupiter with the eastern side lit up.

The planes of the orbits of the inner satellites are tipped to the plane of Jupiter's orbit about the sun by 3°. This means that the sun, as seen from Callisto, sometimes passes just above or just below Jupiter as it crosses the sky. Then there is no eclipse, though Jupiter passes through its new-phase with a crescent of light too thin to be seen and with its atmosphere alight around the rim. The ring of light, brighter on the side toward the sun, will seem washed out and unimpressive, however, in the glare of the sun's light.

On the inner satellites that are closer to Jupiter than Callisto is, Jupiter appears so large that, despite the tipping of the orbits, the sun passes behind it every time. Sometimes it passes behind it above or below Jupiter's mid-line, but it never misses altogether.

If you imagine yourself on any of the inner satellites at a spot on the surface directly under Jupiter, then Jupiter is at the zenith in each case. What you will see will be much the same as what you do on Callisto. The sun (appearing the same size from each of the satellites) will move from east to west, passing behind Jupiter, which will go from half-Jupiter to new-Jupiter to half-Jupiter again. And it will be full-Jupiter at midnight.

Of course, the closer the satellite is to Jupiter, the faster the sun moves across its sky (see Table 46 on page 148), and the more rapidly it moves in its passage behind Jupiter. On the other hand, Jupiter looks larger and larger as one moves from satellite to satellite toward it, and that tends to increase the length of the passage. Taking both factors into account, Table 50 shows how long the eclipse lasts on each of the inner satellites of Jupiter.

There's no question but that the sight is most magnificent on Amalthea. An absolutely gigantic Jupiter, over ninety times as wide as our full moon, would go through its change of phases with enormous rapidity. In three hours after sunset it would go from half-Jupiter to full-Jupiter, and in another three hours from full-Jupiter back to the other half-Jupiter, for the sun would be rising again. We could actually see Jupiter expand and contract as we watched.

And during the eclipse, what a huge circle of sunlight we would see—sunlight shining through the edges of Jupiter's atmosphere. Within that huge circle would be no stars, either, remember.

And how magnificent full-Jupiter would look as seen from Amalthea! Not only would it be a huge circle of solid light, but, as we shall see later, that circle is marked with colored clouds forming belts, zones, and spots, whose turbulence we might be able to see with the unaided eye.

TABLE 50

Eclipses of the Sun as Seen From the Satellites

SATELLITE	MAXIMUM LENGTH OF ECLIPSE (HOURS)	MINIMUM TIME BETWEEN ECLIPSES (HOURS)
Callisto	4.6	396
Ganymede	3.5	165
Europa	2.8	83
Io	2.2	40
Amalthea	1.5	10.5

Nor would full-Jupiter be too bright to look at, even from Amalthea. Though Jupiter is $\frac{1}{14}$ as bright as the sun is (when seen from Amalthea), that brightness is spread evenly over Jupiter's giant disc. Each small piece of gleaming Jupiter, as seen from Amalthea, would be considerably less bright than an equal size piece of the moon as seen from Earth. There would be no trouble at all looking at Jupiter.

One more point. If we stand at different places of the surface of an inner satellite, then Jupiter is in a different part of the sky and the details of the sight vary. Some variations might be very interesting.

Suppose, for instance, that you are standing on a point on Amalthea's surface from which Jupiter is seen in the east, with the horizon cutting it in half. When the sun is setting in the west, Jupiter is in the full-phase. The half above the horizon is completely lit up and has been ever since noon, three hours before. It doesn't look as impressive as it might, though, because the sun is still in the sky.

After the sun sets, the visible portion of Jupiter begins to shrink. Three hours after sunset, when it is midnight at the place where you are standing, the giant planet is in the half-

Jupiter stage, with the half below the horizon in sunlight. Above the horizon you can see only darkness, a semicircular patch in which no star shines.

This continues for some two hours, and then something begins to happen. Very dimly, a semicircle of light begins to appear at the eastern horizon. The sun is about to begin its passage behind Jupiter and does so below the horizon where you can't see it; but you can see its light beginning to shine through the fringes of Jupiter's atmosphere.

The semicircle brightens steadily for an hour and a half until the sun appears triumphantly above Jupiter. As it continues to rise, light begins to spread over the half of Jupiter we can see.

The appearance of the semicircle of light so unexpectedly (unless you knew what was happening) would be exciting indeed the first time you saw it.

Satellites in Callisto's Sky

But let us get back to Callisto at midnight again, and imagine ourselves looking at full-Jupiter at zenith again. Is Jupiter all we would see?

Not at all. We would also see other satellites.

We couldn't see all of them, of course—at least, not with the unaided eye. The small outer satellites are too distant and too small to be seen from any of the inner planets. Amalthea is larger than the outer satellites and is closer, but is still too small to be much of anything. It is, in fact, only the Galilean satellites that stand out in the satellite sky; and this is quite enough, for from Callisto we can see each of the other three: Io, Europa, and Ganymede.

The orbit of each of these three is closer to Jupiter than Callisto's orbit is. This means that all three of them, as seen from Callisto, shift from one side of Jupiter to the other, never departing farther from Jupiter than some fixed maximum amount.

Consider Ganymede, for instance. From Callisto, it will be seen moving from west to east in front of Jupiter, and then from east to west behind Jupiter. Both on the east and west, Ganymede moves only about 35° away from the center of Jupiter. If Jupiter is at zenith, Ganymede moves only a little over a third of the way to the horizon in either direction. Ganymede never sets, then, but remains always in the sky, shuttling back and forth around Jupiter.

Of course if we were on the opposite side of Callisto, we would never see Ganymede in the sky, or the other Galilean satellites either. They would always remain below the horizon. If we were at a spot where Jupiter was always near the eastern or western horizon, Ganymede would set, then rise again at the same spot. If Jupiter were just below the eastern or western horizon, Ganymede would rise as though from nowhere, climb some way up into the sky, and then turn about and sink toward its setting again.

But let's remain with the simple situation in which Jupiter is directly overhead.

In this photograph of Jupiter, the satellite Ganymede (Jupiter's largest) is seen in the lower left. The shadow, seen as a dark spot near the bottom of Jupiter's disc, precedes the satellite when the two are seen from this angle.

Ganymede goes around Jupiter in 7.15 days, but Callisto follows it in the same direction, traveling about Jupiter in 16.69 days. If they start together, both on the same side of Jupiter, then by the time Ganymede has made one complete circle about Jupiter, Callisto has gone on to complete nearly half its orbit. It takes an additional five days before Ganymede catches up and both are on the same side of Jupiter again.

As seen from Callisto, then, Ganymede passes from one side of Jupiter to the other, and then back again, in 12.5 days.

The same argument applies to Europa and to Io. Since they are closer to Jupiter and move faster, they make their circle in less time and then catch up to Callisto in less time. As seen from Callisto, Europa makes the trip from west to east and back to west in 4.5 days, while Io does it in 2 days.

The Galilean satellites change in apparent size as they make their circle. When Ganymede, for instance, is at the extreme west of its shuttle, it is about 900,000 miles from Callisto and has a diameter of 12', a little more than a third the apparent diameter of our moon. As Ganymede moves eastward, it swings between Jupiter and Callisto, so that by the time it is about to move in front of Jupiter's globe, it is only 500,000 miles from Callisto and its diameter has increased to 22'. Then as Ganymede passes Jupiter and goes on eastward, it swings away again and is back to a diameter of 12' when it is at its eastward extreme.

When Ganymede begins heading westward again, it continues to move away from Callisto as it heads to the other side of Jupiter. It continues to shrink in diameter until it is 1,840,000 miles from Callisto, when it is about to pass behind Jupiter and be eclipsed. It is only 6' in diameter then and can just barely be recognized as a tiny globe.

Once it emerges from behind Jupiter and continues westward, it begins to increase in apparent size again until it is back to 12' at the extreme west. Then it all starts over again.

Europa and Io perform similar evolutions as seen from Callisto. Since they are nearer to Jupiter than Ganymede is, they remain closer to Jupiter in Callisto's sky. Europa never moves farther than 21° from Jupiter's center; Io, never more than 12°.

The orbits of Europa and Io are smaller than that of Ganymede. They are not as close to Callisto as Ganymede is when all are on the same side of Jupiter. They are not as far from Callisto as Ganymede is when all are on the opposite side of Jupiter. This means that Europa and Io don't change in apparent size as much as Ganymede does in the course of their orbits.

Of the two satellites, Io is farther from Callisto (when both are on the same side of Jupiter) than Europa is. On the other hand, Io is larger than Europa. The two, therefore, appear to be almost the same size as seen from Callisto, averaging about 7′ of arc. Both get slightly larger and slightly smaller in the course of the orbit, Europa more so than Io; but in neither case is the difference very noticeable.

The satellites, like Jupiter itself, show phases. In fact, each one of the three satellites in Callisto's sky would show, when it was close to Jupiter, the same phase that Jupiter itself showed.

This introduces another change in the pattern. When Ganymede is passing in front of Jupiter, both may be in the full-phase if the passage takes place when the sun happens to be on the other side of Callisto. Both Ganymede and Jupiter can be thin crescents if the sun happens to be near Jupiter in the sky at the moment of passage. Or both can be anything in between.

A particularly interesting sight might be when Ganymede passes in front of Jupiter when both are in the half-phase, especially if you are viewing it from a spot on Callisto where the sun is below the horizon at the time. During half the passage, Ganymede will be moving across that half of Jupiter's surface which is dark. It will then be seen against a small patch of background without stars, like a tiny imitation of its giant primary, and will be seen more clearly than at any other time perhaps.

What about the satellites and the sun? Can a satellite eclipse the sun as seen from Callisto? Well, the sun as seen from Callisto is only 6′ in diameter, which means it can be covered completely by any of the three Galilean satellites in Callisto's sky. At the proper moment, it might even be covered exactly by any of the three, so that the sun's corona might be dimly seen and the effect would be like a miniature imitation of a solar eclipse seen from Earth.

Except, alas, that this never happens.

The orbits of the satellites are tipped 3.1° to the orbit of Jupiter. This means that the sun always passes the satellites a little above or a little below. The only time a satellite would pass in front of the sun (as seen from Callisto or any of the other inner satellites) would be when the satellite and the sun were both behind Jupiter, or when Jupiter was exactly between the two. In either case, no eclipse could be seen.

Could one satellite eclipse another? That is quite another thing. The satellites all move in almost the same plane, so that Io, Europa, and Ganymede, as seen from Callisto, would all be moving back and forth in almost the same line. Unfortunately, Callisto's orbit is the most tipped, so that from Callisto one is often looking a little down or a little up at the other orbits. This means that the satellites move above or below in passing each other.

Sometimes, though, Callisto will be in a spot where it can see one satellite move in front of another. This could be done at any phase so that there would be every kind of pattern to the eclipse. (From the other Galilean satellites, whose orbits are almost exactly in line, the eclipse of one satellite by another would be a more common sight than from Callisto.)

And couldn't a satellite move between the sun and Jupiter? Yes, it could, and it does. Each satellite moves between the sun and Jupiter once in every revolution. Each one casts a shadow, and when it passes between the sun and Jupiter the shadow extends in the direction of Jupiter and can fall upon its face. Naturally, the shadow doesn't cover all of giant Jupiter; it only covers a small circular spot on its surface, a spot a couple of thousand miles across.

When Ganymede, for instance, moves in front of Jupiter (as seen from Callisto), its shadow extends toward Jupiter if both are in the full-phase. Then the sun is shining directly on both Ganymede and Jupiter, and Ganymede's shadow, which is always away from the sun, is toward Jupiter.

If both Ganymede and Jupiter are exactly in the full-phase as seen from Callisto, the shadow of Ganymede extends directly backward. Ganymede then covers its shadow.

If, however, Ganymede crosses Jupiter when both are a little less than full or a little past full, then the sun hits Ganymede at a slight angle and the shadow falls on Jupiter a little behind Ganymede as it moves across, or a little ahead. The shadow, a little dark circular patch, can then be seen to move across the lighted surface of Jupiter; and it can be seen with the unaided eye. (We can see the shadows of the satellites as they cross Jupiter's lighted portion from Earth—using a telescope, of course.)

Sometimes, when Jupiter is full, there is a shadow upon it, even though there is no satellite in the sky to cast it. It crosses Jupiter sometimes quite high up or quite low down, and sometimes it doesn't cross Jupiter at all. Presumably, it misses Jupiter altogether at those times. Often, though, it will cross Jupiter in its full-phase.

It is, of course, the shadow of Callisto itself.

With all this in mind, we can see that the side of Callisto which faces Jupiter is treated to a celestial extravaganza far and away beyond anything that can be seen in Earth's sky.

Remember that Callisto is the Galilean satellite farthest from Jupiter and therefore the easiest and least expensive to reach (since it would require least rocket fuel). What a spot it might be for tourists someday!

The Other Satellite Skies

Shifting now from Callisto to Ganymede, we don't have to go into so much detail, for some things remain the same.

Suppose we stand on that spot on Ganymede's surface that puts Jupiter at the zenith. Jupiter is larger than it appears to be from Callisto, and the sun makes the circuit of the sky in less than half the time it does as seen from Callisto, but the eclipses and the phases stay the same.

From Ganymede, Europa and Io would be seen shuttling back and forth across Jupiter and back, and these two would present no surprises. Callisto, however, being farther from Jupiter than Ganymede is, can never move between Ganymede and Jupiter.

From Ganymede, we can see Callisto move behind Jupiter, going from east to west. As it passes Jupiter on its westward journey, it keeps on going until it sets. It moves back to the other side only by passing behind Ganymede. When that is done, it rises in the east and again moves behind Jupiter.

This means that if you are standing on the side of Ganymede directly away from Jupiter, you will have a view of Callisto every once in a while. Callisto will rise in the east, travel across the sky, and set in the west.

From the side of Ganymede away from Jupiter, something can therefore be seen that could never be seen on Callisto— the occasional eclipse of the sun by a satellite.

Every once in a while, Callisto will get exactly between Ganymede and the sun. At that time Callisto is as close to Ganymede as it can get—about 500,000 miles away. At that moment it is about 22' across, or over three times the apparent width of the sun. As Callisto crosses before the sun, it can hide it completely for up to twelve minutes, and its shadow will cover much of Ganymede.

As seen from Callisto, this would mean that Callisto's shadow, as it fell upon Jupiter, might happen to get to the very center of Jupiter's circle of light (when it was in the full-phase) just as Ganymede, crossing in front of Jupiter, got there. In that case, Callisto's shadow would fall on it.

If we were standing on Europa, with Jupiter at zenith, then only Io would shuttle back and forth from one side of Jupiter to the other. Ganymede and Callisto would both appear in the sky of the side away from Jupiter.

If we were standing on Io, then Europa, Ganymede, and Callisto would all three appear in the sky of the side away from Jupiter.

From Io, however, it would be possible to see Amalthea more clearly than from any of the other Galilean satellites. When Amalthea is between Jupiter and Io, and is therefore at its closest approach to Io, it would have an apparent diameter of 3'. This wouldn't be enough to show it as a distinct globe, and it would look like a bright star. Yet it would be a special kind of star that could attract considerable notice from someone standing on the Jupiter side of Io.

From that side, the three large satellites—Europa, Ganymede, and Callisto—would only be seen when they were heading for the far passage behind Jupiter. Europa, the Galilean satellite nearest to Io, would be 28' in diameter as it rose in the east, and would look almost as large as our own moon. As it moved westward, however, it would move farther from Io and would shrink in apparent size until it was only 18' across as it passed behind Jupiter. Ganymede and Callisto would be somewhat smaller in appearance, and all three, it would seem, would be infinitely more impressive than starlike Amalthea.

But none of three Galileans would pass in front of Jupiter as seen from Io, and Amalthea would.

As seen from Io, Amalthea would never be farther than 25° from Jupiter in either direction. Amalthea would make the sweep from west to east and back again in 16.4 hours.

If Amalthea should happen to pass Jupiter when the planet was in the full-phase, Amalthea would also be full and would be at its brightest, but it would be lost against Jupiter's glare. If Amalthea should happen to pass Jupiter when both were crescents, the satellite would be very dim since so little of it would be lit up and the presence of the sun, nearby in the sky, would be distracting.

Perhaps the best situation would be when Jupiter and Amalthea were seen in the half-phase from some position on Io's surface where the sun was below the horizon at the time. Amalthea, still a fairly bright star, would pass before the darkened part of Jupiter's disc, a semicircular area over seven hundred times the area of our own full moon.

Against this dark starless area, a large moon with a disc would be noticeable enough, but it would be noticeable at any time. A small satellite, however, looking like a bright star, might go unnoticed ordinarily—but not when it crossed that dark area where no other star existed, moving at an almost visible speed. It would then be an outstanding sight and one a tourist might well wait for.

But what if we were on Amalthea itself? There the sights might be most spectacular of all. On that innermost satellite, all four of the Galilean satellites would be visible as discs. It is the only satellite in the Jupiter system, the only solid ground on which an astronaut might stand, from which *four* large satellites could be seen.

Because of Amalthea's innermost position, all four Galilean satellites would be seen in the sky from that portion of the surface away from Jupiter. All would rise in the east and set in the west and all would move across the sky rather quickly—not so much because of their own motion as because of that of Amalthea.

Amalthea whips about Jupiter in 12 hours, so that it is continually overtaking the Galilean satellites, which must move through longer orbits at slower speeds.

Callisto, the farthest of the Galileans, moves most slowly. By the time Amalthea has completed a revolution in 12 hours, Callisto hasn't had time to move much, and in another twelve minutes Amalthea has caught up with it in its new position. For that reason, Callisto makes a complete circle of Amalthea's sky in 12.2 hours.

Ganymede moves faster than Callisto, and it takes a little longer for Amalthea to make one revolution and catch up with it again; still longer for Europa; and longest for Io. Ganymede circles Amalthea's sky in 12.7 hours, Europa in 13.7 hours, and Io in 16.4 hours.

Io, which is the satellite closest to Amalthea, would be 53' across when it was at zenith as seen from that side of Amalthea away from Jupiter. It would be twice as wide as our full moon. At either horizon it would be considerably farther from Amalthea and would be only 34' across, about as large as our moon.

That would make an interesting change and something we are unaccustomed to on Earth. Imagine a glowing circle of light that gets steadily and visibly larger as it climbs to the zenith, and then shrinks again as it moves to the other horizon.

The other satellites would be smaller and would change less in size from zenith to horizon. Europa and Ganymede would both be 17' at the horizon and both would swell at zenith, to 23' in Europa's case and 20' in Ganymede's. Callisto, farthest from Amalthea, would be the smallest: 9' at the horizon and 10' at zenith.

The satellites won't seem as bright as our moon—not even Io when it is at its largest—because they are over five times as far away from the sun as our moon is, and therefore receive less sunlight. However, by shining more softly, they may have even more beautiful shadow effects on their surfaces.

All the satellites move in nearly the same track. As the more distant satellites are the more rapidly moving in Amalthea's sky, they overtake the nearer satellites and pass close by. Occasionally, the farther slips behind the nearer and there is an eclipse of one satellite by another.

Of course, the sun is also moving in the sky of Amalthea, making a complete circle in 12 hours and overtaking each satellite in turn. The pattern of movement of the four satellites is made even more complicated, then, by the fact that they are constantly changing phases.

What's more, different satellites may have markedly different phases while both are in the sky, depending on the position of the sun. If the sun has just set in the west, Ganymede near the eastern horizon will be full, while Europa near the western horizon will be crescent, and Io and Callisto, high in the sky, may be in the half-phase.

There is a chance that the sun may slip behind one satellite or another, and all four are large enough, as seen from Amalthea, to eclipse it completely. However, the track of the sun is tipped to the track of the satellites by 3.1°. The sun can slip behind a satellite and be eclipsed only if it is quite near the point where the two tracks cross and if a satellite happens to be there too.

If we are standing at the surface of Amalthea at a point directly opposite from the center of Jupiter, the crossing will take place just at the zenith. If the sun and a satellite reach the zenith simultaneously, there will be an eclipse of the sun. Naturally, it doesn't happen often.

Still another complication in Amalthea's sky is this. Every twelve hours, Amalthea passes into Jupiter's shadow and stays there for an hour and a half. From the side of Amalthea away from Jupiter, neither Jupiter nor the sun can be seen, so the shadow might go unnoticed except that it could affect the Galilean satellites. Any satellites that happened to be near the zenith at the time and in the full-phase would also pass through Jupiter's shadow, and their light would gradually wink out.

In fact, whenever a Galilean satellite was in the full-phase, no matter where it was in the sky, it would move into eclipse and grow dark (though Callisto might occasionally escape).

On the whole, if we consider the four Galilean satellites and the sun, all in the sky as seen from that part of Amalthea away from Jupiter—if we consider them racing rapidly across the sky, the satellites swelling and shrinking and changing phases in various patterns, covering and uncovering each other and occasionally covering and uncovering the sun, and gradually darkening whenever they reach the full phase—we can fairly say that the sky of Amalthea is the most complicated and fascinating in the entire solar system.

And yet, given their choice, tourists might not wish to stay on the side away from Jupiter where all this can be so clearly seen. They may prefer the side toward Jupiter.

From the side toward Jupiter, the Galilean satellites are much farther away from Amalthea and are smaller and less impressive as they move behind Jupiter on the east and emerge on the west.

But who would look at the satellites, when one could look at giant Jupiter, 46° across? On no solid ground anywhere in the solar system could there be a spectacle so huge, so massive, so frightening as the sight of swollen Jupiter.

However fascinating the skies of Amalthea, whether away from Jupiter or toward Jupiter, there is this disheartening fact to remember. Thanks to the strong gravitational pull of giant Jupiter, Amalthea is the most difficult of the satellites to reach and to escape from. Even after men have managed to reach the Jovian system and to land on some of the satellites and establish bases there, it may still be a long time before anyone ventures quite as close to Jupiter as Amalthea is.

But when the time comes that men can do so, they will find the effort worth it.

8 ≣

THE
ATMOSPHERE
OF
JUPITER

Planets and Clouds

Now that we have completed our trip through Jupiter's satellite system, we must return to Jupiter itself. Having stood, in imagination, on each of the satellites in turn, can we now imagine ourselves standing on Jupiter and observing the heavens?

If we could, we would see much the same kind of sky we would have observed from Amalthea on the side away from Jupiter (for, of course, if we stood on Jupiter's surface, Jupiter itself would not be in the sky). The four large satellites would be in Jupiter's sky, each a little smaller than they would be as seen from Amalthea. Io would look about like our moon in size, Europa and Ganymede would be half our moon's diameter, and Callisto would be a third its diameter. Amalthea would be visible as a bright star.

Since Jupiter rotates in ten hours, it overtakes the motion of the various satellites, and just as on Amalthea, the satellites rise in the east and set in the west.

Callisto, the farthest from Jupiter of the large satellites, is most quickly overtaken since it moves most slowly. It would make a complete circuit of Jupiter's sky in 10.25 hours, so that it would set only a little over 5 hours after it rose. Ganymede would make a complete circuit in 10.7 hours, Europa in 11.3

hours, and Io in 13 hours. Only Amalthea would linger. Since it circles Jupiter in 12 hours, it almost keeps up with Jupiter's surface, which rotates in 10 hours. If you are standing on Jupiter's surface, Amalthea would fall behind so slowly as to take 60 hours to make its circuit. After rising in the east, it would set in the west one and a quarter Earth-days later.

The pattern of eclipses of the satellites and of the sun is much the same as on Amalthea.

But if we stood on the solid surface of Jupiter (or had some detecting instrument on the surface, since we ourselves could not for long stand the gravitational pull) could we really see the changing pattern of the skies? Or would something get in our way?

In order to answer that question, let's begin by considering what we see when we look at the surface of a heavenly body.

Our moon, for instance, shows very clear surface features through the telescope. There are mountains, craters, flat dark areas, and so on. These are never obscured; they are always completely visible. This is because there is no atmosphere on the moon, and we can be sure, therefore, that anyone standing on the moon could see the sky clearly at all times. There would never be any mist or cloud or fog to obscure the sky.

In other words, it works both ways. If we can see the solid surface of a planet clearly because no clouds on that planet obscure our vision, the vision of someone standing on that surface will not be obscured either. He will see out as clearly as we see in.

Mercury, when looked at through a telescope, also shows surface features, but these are much fuzzier and less clear than those on the moon. This is not because they are obscured, however, for we have every reason to think there is no more atmosphere on Mercury than on the moon.

Mercury, however, is considerably farther than the moon. Even at its closest approach to us, Mercury is 200 times as far from us as the moon is. What's more, when it is at its closest, it is very close to the sun, and even if we could see it in the sun's neighborhood, all we would see would be that side of the planet turned away from the sun and sunk in darkness.

When we see Mercury most clearly, it is to one side of the sun and is 400 times as far from us as the moon is. And then, even when it is to one side of the sun, it isn't so far that the sun's brilliance doesn't interfere. All in all, it isn't surprising that we see Mercury's surface so poorly, but that is no reason to suppose that someone standing on its night-surface could not see the starry sky perfectly clearly.

Jupiter's satellites are in the same position Mercury is. They are also thought to have no atmosphere, yet sheer distance hides their surface. Jupiter's satellites, at their closest, are 1600 times as far from us as the moon is. We can make out some vague surface features on those satellites through our telescopes, but no detail to speak of.

What about Mars, though? It approaches us more closely than Mercury ever does. At its closest it is only 140 times as far away as the moon is. What's more, in the case of Mars, we see its sunside when it is closest, so that we see a "full-Mars."

It is not surprising, then, that we can see features on Mars, more clearly than on any other heavenly body but the moon. And yet we don't see them as clearly as we might expect to. The Martian features are fuzzier than would seem likely from distance alone. The reason for that is that Mars has an atmosphere. The atmosphere is much thinner than Earth's atmosphere, only one-hundredth as thick, yet even so it interferes.

The gases of an atmosphere absorb some light, and force light rays into following a slightly curved path because of temperature differences from spot to spot in the atmosphere. The bending of light rays seems to make a particular spot on the surface move slightly from place to place so that it looks blurry and out of focus. (It is such bending of light rays in our own atmosphere that makes the stars move slightly from spot to spot and seem to twinkle.)

Once in a while there are clouds in the Martian atmosphere, and these help obscure parts of the surface. This hardly ever happens since Mars is so dry, but sometimes the thin wind kicks up a sandstorm that will block our view completely. (Rocket probes that have passed near Mars have sent back pictures of the surface far more clearly than anything we could achieve from Earth itself—but that's another matter.)

Again, what works looking in also works looking out. Anyone standing on the surface of Mars would be able to see the skies, but not quite as clearly as he could on the moon. And if a sandstorm arose which obscured the Martian surface to us, it would obscure the Martian sky to an observer on that surface.

Even so, an astronomer on Mars would be more fortunate than his counterpart on Earth, and would be able to count on seeing the sky better, on the average.

The Earth has an atmosphere that is much thicker than that of Mars. What's more, Earth is a watery planet with vast oceans, so that our atmosphere is frequently full of clouds of water droplets. If Earth is viewed from the surface of the moon, its surface is covered with curlicues of white clouds, which obscure much of the land and sea to any outside observer.

To be sure, the clouds don't obscure all of Earth's surface all of the time. A glimpse of various parts of the ocean surface and land surface can be seen now and then. Certain desert areas (including northern Africa, Arabia, central Australia, southwestern North America, and so on) are hardly ever covered by clouds and can be seen clearly. Other places can be seen only occasionally, but any patient watcher on the moon would eventually see enough to be able to get a pretty good idea of the oceans and continents.

The difficulty of seeing through Earth's clouds from outside is an indicator of the difficulty of seeing through them from the inside. Astronomers trying to study the heavenly bodies from Earth's surface are frequently balked by clouds, fog, and mist. To minimize this, astronomers try to build their observatories in deserts where there are few clouds or on mountaintops where the air is thin.

But what if a planetary atmosphere is even thicker and cloudier than Earth's is? Consider Venus, for instance.

Venus is the planet closest to us, sometimes only 100 times as far from us as the moon is. Like Mercury, though, it circles the sun more closely than we do, so that when it is at its closest we see only its night side. Still, Venus is larger than Mercury, and besides being closer to us, it can be seen farther from the disturbing brilliance of the sun. Because of all this, we might expect to see the features of the surface of Venus more clearly than those of the surface of Mercury.

But we can't. In fact, we can see no features on Venus at all. What we see, when we look at Venus, is an unbroken whiteness covering that part of the sphere that is lit by the sun.

The trouble is that Venus has a thicker atmosphere than Earth has, and what seems like its surface as we look at the planet is no true solid surface at all. It is a continuous and permanent cloud cover that bars any view of any part of the solid surface beneath.

As it happens, astronomers have shot beams of radar waves at Venus. (Radar waves are like light waves but are many times longer.) The long radar waves can penetrate the cloud layer, which light waves can't, and can reach the solid surface of Venus lying below the clouds. The radar waves are reflected from the surface of Venus, penetrate once again through the clouds, and reach the Earth once more. From the changes imposed on the radar waves in the course of the reflection, astronomers can deduce certain properties of the solid surface of Venus. They have been able to detect mountain ranges, for instance.

Again, it works both ways. If someone could stand on the

surface of Venus and look up to see the sky, he could no more see out past the clouds than we can see in. Unless he used radar, he would see only the permanent dark gray of the cloud cover. On Earth we see the clouds of Venus as white, because we see them by the light they reflect, which is a great deal. Someone on the surface of Venus would see the clouds as dark gray because he would see them only by the light that penetrates through them, which is very little.

Well, then, if we are to consider how much of the starry sky can be seen from Jupiter's surface, we must consider what it is we see when we look at Jupiter.

Belts and Spots

Jupiter is as far from us as its satellites are, of course, and if we can see no surface features on the satellites to speak of, ought we to expect to see any features on the surface of Jupiter?

Yes, we might. Jupiter is, after all, far larger than its satellites, thirty times as wide across as even its largest satellite, and we might see large features on such a globe even if we couldn't make out any tiny details.

At first, of course, nothing was seen. In the very primitive telescopes of Galileo, Jupiter just appeared to be a small featureless ball of light.

The first report of anything more came in 1659, a half-century after Galileo had turned his telescope on the sky. In that year Christian Huygens published a book in which his drawing of Jupiter, as it appeared in his telescope, showed two straight streaks crossing its mid-portion.

By 1663, Cassini had telescopes which could make out Jupiter's surface still more clearly. He could see the streaks well enough to make out certain irregularities along their lengths. Studying these irregularities, he noticed that they moved along the surface of the planet from the western edge to the eastern.

They disappeared behind the eastern edge and eventually appeared at the western again.

It seemed clear to Cassini that Jupiter was rotating about its axis from west to east as Earth does. By measuring the length of time it took for some particular irregularity to make its way from west to east and back to west, he could calculate the time it took Jupiter to make one rotation about its axis.

Cassini could also show from the direction of movement of these irregularities that Jupiter's axis was only slightly tipped with respect to the plane of its revolution about the sun, and that the streaks themselves were parallel to Jupiter's equator.

The most conspicuous marking is one just north of Jupiter's equator. It is wide and usually dark and is called the "North Equatorial Belt." South of the equator is a similar marking, which is less prominent partly because it seems double with a light streak down its center. This is the "South Equatorial Belt." In 1951, the South Equatorial Belt was so pale as to be almost invisible, but by 1956 it had darkened to the point where it was almost as prominent as the North Equatorial Belt. It is these two belts which Huygens and Cassini noted and studied three centuries ago.

Better telescopes revealed fainter and more delicate belts lying beyond the prominent equatorial belts. There are two to the north, the "North Temperate Belt" and the "North North Temperate Belt" and, of course, two to the south, the "South Temperate Belt" and the "South South Temperate Belt." The polar regions are also dark. In between the belts are lighter areas called "zones."

The over-all color of Jupiter is yellow, but the belts can be anything from a slightly darker yellow to a deep brown. Occasionally, blue and gray tinges may appear.

The belts are not perfectly uniform in their stretch about the planet, but vary in width from place to place. In addition, as Cassini had noticed in the beginning, spots can appear along their lengths, and persist for shorter or longer periods. Sometimes these spots are lighter than the belt itself, sometimes darker.

In short, Jupiter's surface, as seen through the telescope, offers an ever-changing panorama.

It is not likely that a solid surface could have such regularly shaped features so ever-changing in detail. It is more likely that what we are seeing when we look at Jupiter is its cloud cover.

This became certain when the periods of rotation about the axis were obtained for objects at different distances from the equator. An irregularity in the equatorial belts or in between moves around Jupiter in 9 hours, 50 minutes, and 30 seconds (9.842 hours). An irregularity beyond either equatorial belt moves around Jupiter in 9 hours, 55 minutes, 41 seconds (9.930 hours).

The difference isn't much, but it is important just the same. If we were seeing the solid surface of Jupiter, we would have to assume that the planet was rotating all in one piece. The period of rotation would be exactly the same no matter where on the surface we measured it.

If, instead, we were observing a cloud layer, it is not at all impossible that some parts might rotate faster than other parts.

We come to the conclusion, then, that we are not looking at Jupiter's solid surface when we see what *seems* to be its surface. We see only clouds, and somewhere below those clouds is the solid surface. And, as on Venus, someone standing on Jupiter's solid surface and looking up could see nothing of the heavenly bodies, but only the undersurface of the cloud layer.

But if we are looking at Jupiter's atmosphere, why the belts? Venus has a thick atmosphere, but its clouds don't form into belts. Venus's clouds are even and completely featureless.

This may be the result of differences in speeds of rotation. Venus rotates very slowly, once in 5,834 hours, so that its surface, even at the equator where it moves most quickly, moves at only 4 miles an hour. Jupiter, on the other hand, rotates in just under 10 hours, so that its equatorial surface whips along at 28,000 miles an hour.

Now consider the differences in speed. Venus's surface moves at speeds ranging from 4 miles an hour at the equator to 0 miles an hour at the poles. Jupiter's surface ranges from 28,000 miles an hour at the equator to 0 miles an hour at the poles.

This difference in speeds would send Jupiter's atmosphere into spirals, if it moved north and south, as takes place on Earth. (Earth, as seen from space, has clouds in spiral shapes, even though the speed of its surface ranges from only 1,000 miles an hour at the equator to 0 miles an hour at the poles.)

Apparently, something colored exists in Jupiter's atmosphere, and that atmosphere spirals in such a way that the colored substances concentrate largely at certain latitudes and spread out to form the belts. Why at certain latitudes and not at others we cannot say—we still find it hard to explain the circulation of Earth's atmosphere and we have no hope as yet of understanding the fine details of Jupiter's atmosphere.

If it were just a matter of rotation, we might expect Jupiter's belts to be perfectly straight and even. As I said earlier, they aren't. There are always little bulges, hollows, and spots appearing from time to time, expanding, contracting, darkening, lightening. Perhaps these are the results of storms on Jupiter which are violent enough to overcome the normal atmospheric belt-producing circulation, and which follow courses dictated not only by Jupiter's rotation but by unevennesses in the solid surface below.

If we could see Jupiter from a closer view, the irregularities in the atmosphere might be seen to be more pronounced and numerous. We might see more of them in greater detail— enough of them and enough of their structure to mask the belt. Perhaps it is only our great distance and inability to see detail that washes out small differences and makes a more or less smooth belt out of an enormously varying panorama.

Occasionally certain irregularities can persist for years, but all of them must take a back seat to a particular marking on Jupiter that is the most amazing and puzzling of all. It is a large, oval-shaped marking which introduces an interruption into the South Equatorial Belt.

It was first noted by the English scientist Robert Hooke, in 1664, and for a while it was called "Hooke's Spot." Cassini began to study it that same year, and in a drawing of Jupiter which he made in 1672, he showed it as a large round spot.

Occasionally, in the centuries that followed, others would show similar objects in their drawings. In 1831 a German astronomer, Heinrich Samuel Schwabe, made a very clear drawing of Jupiter showing the spot.

It is only looking back on it now, though, that we find these early reports of interest. Prior to a century ago, the reports were only sporadic and it was assumed that the spot was only another of the temporary features of Jupiter, no more important than the others.

The change came in 1878, when the spot was observed by another German astronomer, Ernst Wilhelm Tempel. He described it as an oval ellipse about 20 degrees south of Jupiter's equator and with its long diameter parallel to the equator. It was distinctly reddish in color and has been called the "Great Red Spot" ever since.

From 1878 on, it has been under constant observation, and it is quite obviously a long-lived feature. Its color changes, and it is more brightly red sometimes than others. In fact, it can become quite pale and stay so for decades. When this happens, it is not very noticeable—especially with a poor telescope, when it might not be seen at all.

Apparently, early astronomers with their poor telescopes saw the Great Red Spot only when its color was dark but not otherwise. The result was that reports were only occasional, and in the periods when it was not seen those reports were forgotten.

Then when the color happened to deepen considerably in the late nineteenth century and a good telescope was turned on Jupiter, its appearance was spectacular enough to place it firmly into the astronomical consciousness.

Nowadays we can see it even when it is pale. We know it is there even if it shows no color at all, for there is an indentation into the South Equatorial Belt into which the Great Red Spot fits, and that indentation remains even when the Spot's color has washed out into near invisibility.

In addition to its apparent permanence, the Great Red Spot is remarkable for its size. The adjective "Great" is not misplaced. It is about 30,000 miles (45,000 kilometers) long and about 8000 miles (13,000 kilometers) wide. The area of the Red Spot is about 175,000,000 square miles, or very nearly that of the total area of Earth.

It is clearly a cloud formation and is not attached to the solid surface of the planet. At times it moves ahead or falls

When Jupiter is photographed in blue light, as in this case, those portions of the surface that reflect little blue light look particularly dark. For that reason, the Great Red Spot (which reflects red light, rather than blue) looks very prominent.

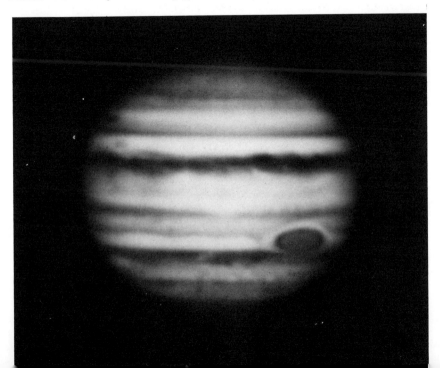

behind the rest of the cloud formations, gaining or losing one entire lap, or even more on occasion. It changes only in longitude, however, never moving north or south of its position 20 degrees south of Jupiter's equator.

What is the Red Spot? Why is it red? Why does the color grow darker and lighter? Why does the Red Spot move about relative to other parts of the planet? Why does it move east and west but not north and south?

Even today, nobody understands the Great Red Spot, though there have been many theories.

One of the early theories that was quite popular toward the end of the nineteenth century was that all the outer planets were still hot. It was thought that the planets had originated from material in the sun and had gradually cooled off. The smaller worlds had cooled fairly quickly and were now quite cold and solid, at least on the surface. (The presence of volcanoes and the molten rock they spewed out was ample evidence that there was still great heat below the surface even on Earth.)

The larger planets held more heat and cooled down more slowly and were therefore (it was thought) still hot, with enormous volcanic activity. The colored belts might exist in regions of greater (or lesser) heat. The Great Red Spot might be the crater of a huge volcano, and we might be looking down into its fires which dimmed and brightened with the level of its activity.

All this, however, turned out not to be so. It is no longer thought that the planets originated directly from the sun, but that they formed, instead, from the conglomeration of dust and gas on the outskirts of the huge whirling mass that condensed to form the sun.

For that reason the planets were not hot at the beginning, but were formed out of cold matter. Of course, they may have grown hot because as the matter came together to form large globes, gravitational forces compressed and heated the matter at the center of the globe. Even Earth has a large core of molten iron, for instance. Furthermore, radioactive elements trapped in the outer crust give off heat slowly as they decay, and that helps keep the crust of the Earth hotter than it would otherwise be and may account, in part at least, for the heat produced by volcanoes.

Jupiter, being a much larger world than Earth, would develop far more pressure and heat at its center and might contain more radioactive materials too. However, astronomer's don't think that any reasonable course of events in the formation of Jupiter would heat it up to the point where the *surface* was red-hot.

What's more, the amount of heat radiated by Jupiter can be actually measured by very delicate instruments, and astronomers can then calculate the temperature of the surface we can see. In 1926 an American astronomer, Donald Howard Menzel, showed that Jupiter's surface temperature is no higher than —135° C.

Of course that temperature is the temperature of the visible cloud layer, which might be expected to be colder than the solid surface beneath. (Our upper atmosphere is considerably colder than Earth's surface is.) However, even allowing for that, the temperature is too low to make it possible for the surface to be red-hot. Astronomers are certain that Jupiter and the other outer planets are cold bodies, and surface markings must be explained with that in mind.

The Oxygen Atmosphere

Well, then, assuming Jupiter to be cold, what can we say about Jupiter's cloud layers? Are they clouds like those on Earth? If so, why are they colored, and why do the colors change? If not, what are the clouds made of?

Let's consider atmospheres in general for a moment, and perhaps we can eventually try to answer that question.

The atoms making up solid matter cling to each other by

"electromagnetic forces" involving electrons—forces that are extremely strong. Atmospheres, however, are made up of gases in which single atoms (or molecules made up of several atoms held together by electromagnetic forces) do not cling to each other but move about independently. They are held to the surface of the planet only by gravitational forces, which are much weaker than the electromagnetic ones.

The atoms or molecules in the atmosphere move about in a random manner, hitting each other blindly and then rebounding. Some of them manage to find their way (by sheer chance) to that portion of the atmosphere hundreds of miles above the planet's surface. The atmosphere gets thinner and thinner with height (the lower layers being compressed by the weight of those above), and when atoms or molecules are high enough, so few of them are present that the likelihood of further collisions among them becomes very small. In that case, an atom or molecule that happens to be moving upward may not strike another until it is far out in space and the possibility of its moving back into the neighborhood of the planet is unlikely.

In other words, every atmosphere has a tendency to leak slowly outward into space.

Naturally, in order for a particular atom or molecule to move upward into space indefinitely, it must be going at more than the escape velocity. The stronger the gravitational field, the higher the escape velocity and the less likely it is that an atom or molecule might be going fast enough to escape.

For an object like the moon, which has a surface gravity only ⅙ that of the Earth, the chance that an atom or molecule might be going at velocities higher than the 1.5 mile-per-second escape velocity is quite good. This means that the atmosphere leaks away rather quickly, so that today the moon lacks an atmosphere altogether. Maybe it never had an atmosphere, lacking the gravitational force to gather the gases in the first place; but even if it once did have one, that atmosphere has long since gone.

For the same reason, other small bodies like Mercury and the moons of Jupiter don't have an atmosphere.

On the other hand Mars, which has a surface gravity 2.5 times that of the moon, has managed to hang on to a thin atmosphere—one that is only $\frac{1}{100}$ as thick as the Earth's, but it is there. The Martian atmosphere leaks too, but the leak is so small that some has remained over all the billions of years that the planet has existed.

Venus has a surface gravity 2 times that of Mars, and Earth has a surface gravity 2.5 times that of Mars; both have fairly dense atmospheres. It is not surprising, then, that the outer planets, with surface gravities as strong as Earth's or stronger, should also have atmospheres that are as dense as Earth's or denser. This would be particularly true of Jupiter, which has a stronger gravitational pull and a higher escape velocity than any other planet in the solar system.

But there are atmospheres and atmospheres. Some may be made up of a set of gases in particular proportions, others of the same set in different proportions. Still others may be made up of different gases altogether.

We know that Earth's atmosphere is made up chiefly of nitrogen and oxygen in a ratio of 4 to 1. There are also present minor quantities of other gases. There is a gas called argon, for instance, which makes up 1 percent of Earth's atmosphere, and there is carbon dioxide which makes up 0.03 percent of Earth's atmosphere.

The most curious thing about Earth's atmosphere is the presence of so much free oxygen. Fully 21 percent of the Earth's atmosphere is oxygen. Oxygen, however, is a very active gas. We would expect it to combine with various chemicals in the Earth's crust and to disappear rather quickly.

Every animal on Earth breathes in oxygen and combines it with carbon in its food to form carbon dioxide. When wood or coal burns, carbon combines with oxygen to form carbon dioxide. Why doesn't all the oxygen in the atmosphere combine with carbon to form carbon dioxide, and why isn't there carbon dioxide in the atmosphere instead of oxygen?

The reason this doesn't happen is that plant life feeds on carbon dioxide, using the energy of sunlight to combine it with water and other substances to form all the materials in the plant structure. In the process, free oxygen is liberated. This action of plant life keeps oxygen in the atmosphere by forming it as fast as it is consumed.

What about atmospheres on planets that don't have life? If the atmospheres were like that on the Earth, they should be rich in carbon dioxide since there would be no life forms to convert it to oxygen.

One way of telling the nature of the atmosphere on other planets is to study the light reflected from them. Different molecules absorb different parts of the sunlight. By noting the "absorption spectrum" of the reflected light with special instruments, it is possible to make some conclusions about the atmospheres of particular planets.

The atmospheres of Mars and Venus happen to be very rich in carbon dioxide. In fact, they are almost entirely carbon dioxide. That alone tells us that there are probably no life forms on those planets, at least none that are in any way similar to plant life on Earth. (Of course, there may be life form of other kinds with other chemistries.)

Ought we to suppose from our experience with Earth and its next-door neighbors in space that all planetary atmospheres are made up of various combinations of gases such as nitrogen, oxygen, and carbon dioxide?

Actually, we can be pretty sure this is not so. By studying the light that reaches us from the sun and from other stars, we discover that some wavelengths of light are missing. There are dark lines in the spectra of those stars. (A spectrum is light that has been spread out into its different wavelengths.) These dark lines represent wavelengths absorbed by the gases near the surface of the star. Each different wavelength can be absorbed only by a certain kind of atom, so that by studying the missing wavelengths, we can determine the kind of atoms present in the stars. It turns out that the average star is mostly hydrogen.

In fact, astronomers who have studied spectra in detail have decided that about 90 percent of all the atoms in the universe are hydrogen. Another 9 percent are helium, and the remaining 1 percent includes all the other varieties of atoms.

TABLE 51

Relative Numbers of Atoms of the Various Elements in the Universe

ELEMENT	NUMBER OF ATOMS (PHOSPHORUS = 1)
Hydrogen	4,000,000
Helium	310,000
Oxygen	2,150
Neon	860
Nitrogen	660
Carbon	350
Silicon	100
Magnesium	91
Iron	60
Sulfur	38
Argon	15
Aluminum	10
Calcium	5
Sodium	4
Nickel	3
Phosphorus	1

Let us suppose, for instance, that we have scooped up a small sample of matter that represents the average composition of the universe. That sample contains 1 phosphorous atom. Since there is five times as much calcium as phosphorus in the universe, we would expect the sample to contain 5 calcium atoms. In similar fashion we can work out the numbers of atoms of other varieties present, and that gives us Table 51.

Only 16 elements are listed out of the 105 that are known. All the others are present in amounts even smaller than phosphorus is. The chances are that not even one atom of these will be present in the sample.

Hydrogen is the simplest atom, so the mass of one hydrogen atom is set, for convenience, at just about 1. Under planetary conditions, hydrogen exists in molecules made up of two hydrogen atoms each. The mass of the hydrogen molecule is therefore 2.

Helium is the second simplest atom. Under planetary conditions, it exists as single atoms and its mass is 4.

The smaller the mass of an atom or molecule, the more rapidly it moves at any particular temperature. At the temperature of the Earth, it is impossible for Earth's gravity to hold the fast-moving hydrogen molecules and helium atoms. It can't hold neon either, which exists as single atoms with a mass of 20.

Oxygen, however, (a two-atom molecule with a mass of 32), nitrogen (a two-atom molecule with a mass of 28), and carbon dioxide (a three-atom molecule—one carbon atom and two oxygen atoms—with a mass of 44) *can* be held by Earth's gravity. It is for this reason that Earth's atmosphere is made up of oxygen and nitrogen. They leak so slowly that such an atmosphere can last for billions of years.

Carbon dioxide, the most massive of these gases, leaks most slowly, and even Mars's small gravity can hold carbon dioxide.

Not all the hydrogen has left the Earth, of course. Some

hydrogen atoms formed combinations with other elements and remained behind. Most of it now is in combination with oxygen to form water (with molecules made up of two hydrogen atoms and an oxygen atom). These molecules make up the vast oceans of the planet.

Helium and neon, however, do not combine with any other elements. They remain *only* in the form of individual atoms, and they are gone altogether. Only tiny traces remain on Earth.

The Hydrogen Atmosphere

But what about Jupiter? Jupiter formed farther from the sun than Earth did, much farther. Its temperature was lower. The lower the temperature, the more slowly the various atoms and molecules move. Even the tiny atoms of hydrogen and helium were slowed down so much that it became rather difficult for them to escape from a gravitational hold.

Hydrogen and helium, numbed by cold, so to speak, didn't leak away from Jupiter as it formed quite as easily as they leaked away from Earth. More hydrogen and helium collected about growing Jupiter, so that it steadily became larger than Earth.

This introduced a "snowball" effect. The larger Jupiter became, the stronger its gravitational force became and the more able it was to hold on to hydrogen and helium, so that it grew still larger—and larger—and larger—like a snowball rolling downhill.

Somewhere between Mars and Jupiter there must have come a turning point, a place where the temperature was low enough to allow sufficient hydrogen and helium to accumulate and reach the point where the snowball effect could begin. That may be why the inner planets are pygmies and the outer planets are giants.

Jupiter is the largest of the giants because it is the nearest to the sun. The closer a planet forms to the sun, the thicker is the dust and gas surrounding it, and as long as that planet isn't so close that the snowball effect won't work, it will grow larger.

But Jupiter grew to its large size chiefly because it could hang on to the hydrogen and helium, while Earth could not; this means that Jupiter and the other Jovian planets must be quite different chemically from Earth and the other Terrestrial planets.

Jupiter's atmosphere must be the product of its chemistry; it must be composed mostly of those two gases, hydrogen and helium. It is not easy to check this, however, because hydrogen and helium (and neon, too) absorb light waves in ways that are difficult to detect. From the reflected light of Jupiter it is hard to tell whether hydrogen, helium, or neon are present and in what proportions.

However, that does not end matters. Perhaps we can make further deductions that *can* be checked. Hydrogen is present in such large quantities that it combines with almost every other type of atom that is present. It doesn't combine with helium or neon, of course, since nothing can combine with those two gases. What about other elements, though?

If we ignore the unsociable helium and neon, then the most common atoms in the universe after hydrogen are oxygen, nitrogen, and carbon. In forming a planet, oxygen, nitrogen, and carbon atoms will all combine with the hydrogen atoms that are present in such overwhelming quantity. Each oxygen atom will combine with two hydrogen atoms to form a water molecule (H_2O). Each nitrogen atom will combine with three hydrogen atoms to form an ammonia molecule (NH_3). Each carbon atom will combine with four hydrogen atoms to form a methane molecule (CH_4).

These compounds can be gases and can make up part of an atmosphere. In Table 52, we have the liquefying temperatures (temperatures below which the substance is a liquid or, in the case of carbon dioxide, a solid) of the various common substances that can be found in atmospheres.

At Earth's temperature, water is a liquid. However, it vaporizes easily so that it is present in the atmosphere in variable quantities as water vapor. Some of this vapor is constantly condensing and forming clouds made up of tiny water drop-

TABLE 52

Liquefying Temperature of Some Substances

SUBSTANCE	LIQUEFYING TEMPERATURE (DEGREES CENTIGRADE)
Helium	− 268.9
Hydrogen	− 252.8
Neon	− 245.9
Nitrogen	− 195.8
Argon	− 185.7
Oxygen	− 183.0
Methane	− 161.5
Carbon Dioxide	− 78.5
Ammonia	− 33.4
Water	+ 100.0

lets or even tiny ice crystals. All the other substances in Table 52 are gases at Earth's temperature.

On Jupiter, with its temperature considerably below that of Earth, water would be permanently in the form of ice. It is not likely that there would be any water vapor in Jupiter's atmosphere.

Ammonia on Jupiter would probably be in the same position that water is on Earth. It would be liquid or solid but easily vaporized. There could be ammonia vapor in Jupiter's atmosphere, and it could settle out in vast clouds. Perhaps the cloud layers we see are largely composed of ammonia.

As for methane, that liquefies at so low a temperature that it should exist as only a gas even at Jupiter's temperatures. We can guess then that Jupiter's atmosphere might be largely hydrogen and helium, plus a little methane, and that it ought to be filled with ammonia vapor and ammonia clouds.

But need we guess? Light coming from Jupiter is sunlight that has penetrated part of Jupiter's atmosphere and has then been reflected. In the course of the penetration, some wavelengths have been absorbed by the molecules in its atmosphere. As it happens, ammonia and methane absorb strongly and should be detectable even if present in the atmosphere in only small quantities. And if they are present, that would indicate a high probability for the presence of hydrogen, helium, and neon as well.

In 1932 a German astronomer, Rupert Wildt, was able to study the reflected light with sufficient detail to show that there were missing wavelengths that definitely placed both ammonia and methane in Jupiter's atmosphere.

Ammonia and methane were also detected in the atmospheres of the other Jovian planets, but the signs of ammonia grew weaker and those of methane grew stronger as farther and farther planets were studied in this way. This is not surprising. The farther the planet from the sun, the colder it is. In the more distant planets the temperature is so low that ammonia freezes solid and very little of it can vaporize into the atmosphere. Methane, which doesn't freeze until extremely low temperatures are reached, remains in the atmosphere even in a planet as distant from the sun as Neptune.

Although the presence of ammonia and methane made it seem certain that hydrogen was also present in Jupiter's atmosphere, direct evidence was eagerly sought.

In 1952 Jupiter was going to pass in front of the star Sigma

Arietis. This event was closely observed by two American astronomers, William Alvin Baum and Arthur Dodd Code. As the star made visual contact with Jupiter's globe, it didn't wink out all at once. First it passed through the thin atmosphere above Jupiter's cloud layer. The starlight gradually dimmed as it was seen through thicker and thicker layers of the atmosphere, until it vanished behind the clouds altogether.

The rate of dimming depended on a number of things, including the mass of the molecules it passed through. It was difficult to keep track of that rate with the instruments available to Baum and Code, but the measurements they made indicated that the average mass of the molecules the starlight passed through was 3.3. This meant that the atmosphere had to be chiefly helium and hydrogen, since no other gases had atomic or molecular weights low enough to account for the 3.3.

The 3.3 figure made it seem there was more helium than hydrogen in Jupiter's atmosphere, but later and still more delicate investigations showed this not to be the case. In 1963 an American astronomer, Hyron Spinrad, studied a light region where hydrogen molecules absorbed strongly enough to be detected, and estimated that the atmosphere of Jupiter was roughly 60 percent hydrogen by weight and 36 percent helium. Also present was 3 percent neon and 1 percent methane. Ammonia vapor would be present in still smaller quantities. Since hydrogen atoms are so light, it takes a great many of them to build up to 60 percent by weight. This means that about 85 percent of all the atoms in Jupiter's atmosphere are hydrogen.

On May 13, 1971, Jupiter passed in front of another star, Beta Scorpii this time. A close study of the way in which its light disappeared, with instruments that were advanced over those of twenty years before, showed that the atmosphere was indeed very largely hydrogen. (The satellite Io also passed in front of the star seven hours later and proved to have no atmosphere.)

In fact, astronomers have grown so adept at studying Jupiter's atmosphere that they can even detect isotopes, or varieties of elements. Hydrogen, for instance, consists chiefly of an isotope called hydrogen-1. It also contains small quantities of a heavier isotope called hydrogen-2 or deuterium. In 1972, American astronomers at the University of Texas detected absorption bands in Jupiter-light which were caused by methane (CH_4) in which one of the hydrogen atoms was of the deuterium variety.

9

THE
PUZZLES
OF
JUPITER

Color and Radio Waves

But if the atmosphere of Jupiter is hydrogen, helium, neon, methane, and ammonia, where do the colors of the bands come from? All these substances are colorless when they are gases or liquids, and white when they are solids.

One possibility, suggested by the English-American chemist Francis Owen Rice in 1955, is that we are dealing with fragments of molecules.

Usually the atoms in molecules are held together by pairs of electrons between them. It is possible to pull away an atom and its electron, leaving behind a fragment of a molecule with an unpaired electron. Such a fragment is called a "radical."

Under Earth-like conditions, radicals do not exist long. They are very active and readily attach themselves to other atoms or molecules. They are then intact again, with all electrons in pairs—and they are then colorless.

On Jupiter, where the atmospheric temperature is lower than on Earth, and where all molecules are more sluggish, radicals attack other atoms or molecules with less vigor and remain in existence longer as "free radicals." Eventually, even on Jupiter, they recombine, but whatever process forms them keeps on forming them, so that there are always enough to lend a color to the atmosphere.

It may be that they are formed in the lower atmosphere, but that the atmospheric circulation whirls them aloft and then concentrates them in certain latitudes, producing the belts. When the rate of free radical formation races ahead of free radical combination, the colors darken; when the rate of free radical formation falls behind, the colors dim. Naturally, we have no way of knowing the details.

Local storms of more than usual intensity may produce the spots in the belts, and these may be pale or dark depending on the quantity of free radicals whipped along by those storms.

The Great Red Spot may be produced by some particular and unique unevenness in the solid surface of Jupiter which, as the atmosphere moves past it, sprays free radicals high in the air. (This alone leaves many difficulties unexplained. Why is the spot so red; is it a different kind of free radical? Why does it move east and west but not north and south? Why is there only one?)

There is also a suggestion advanced in 1970 by astronomers at Massachusetts Institute of Technology, that the color may be due not to free radicals but to sulfur. Sulfur is common enough to be present in Jupiter's atmosphere in small quantities in combination with two hydrogen atoms for each sulfur atom: hydrogen sulfide (H_2S).

Hydrogen sulfide is itself colorless, but free sulfur is yellow and there are many colored sulfur-containing molecules. It is possible to dream up ways in which such colored molecules can be formed from hydrogen sulfide under the conditions in Jupiter's atmosphere.

Whatever the cause of the colors, whether free radicals or sulfur compounds, it takes energy. Without energy, the atoms and molecules in Jupiter's atmosphere would remain in the simplest and least energy-containing forms—which are all colorless.

Where does the energy come from? The obvious and almost

only answer is the sun. Even though the sun's radiation, at the distance of Jupiter, is only 1/25 the intensity of that at the Earth, it is enough to keep Jupiter's atmosphere in constant turbulence. (Saturn, farther from the sun than Jupiter is, has an atmosphere that is far less turbulent and far less colored.)

It seems unlikely, though, that the sun's energy is the direct cause of the formation of colored compounds. If it were, would not the entire atmosphere of Jupiter be colored? Rather, the colored substances would seem to be formed in the lower atmosphere and to be whirled aloft by winds. The nature of the circulation of Jupiter's atmosphere would then concentrate the colored compounds in certain latitudes and form the belts.

But where is the energy source in the lower atmosphere? If the sun's energy sets up turbulence in the atmosphere, that turbulence may produce thunderstorms with a release of intense energy in the form of lightning. This happens in Earth's atmosphere, and why should it not in Jupiter's far vaster and thicker atmosphere?

To be sure, we can see no signs of lightning in Jupiter's atmosphere. Any lightning that is produced would be well below the cloud layer. But lightning liberates more than light; it liberates radio waves too, and some of these can penetrate a cloud cover. Is it reasonable, then, to search for radio waves originating in Jupiter's atmosphere?

Actually, every object that exists is above absolute zero in temperature, and every object which is above absolute zero in temperature emits radio waves. For most ordinary objects the emission is too feeble to detect, but whole planets are another matter, especially when there is a turbulent atmosphere that puts out unusually large quantities.

It was not till after World War II that astronomers developed efficient radio telescopes which were capable not only of detecting radio waves from space but (which is much more difficult) of pinpointing their origin.

In 1955 two American astronomers, Kenneth Linn Franklin and B. F. Burke, studying radio waves which had been puzzling observers for several years, finally managed to demonstrate that Jupiter was the point of origin. Since then, the radio waves coming from Jupiter have been studied intensely.

Some of them are just the kind of radio waves you would expect to be emitted by an object at the temperature of Jupiter. There are other radio waves which are so energetic, however, that if they originated simply because of the temperature of the radiating object, we would have to expect Jupiter to be as hot as the sun. Clearly, these radio waves must originate out of something other than mere temperature. Some of them could be emitted by lightning flashes a billion times as powerful as those liberated in Earth's atmosphere. It seems reasonable to suppose, then, that it may be thunderstorms on Jupiter that produce both radio waves and the planet's colored belts.

Some of the radio waves emitted by Jupiter seem to be of a kind that requires still another explanation. The answer came from studies here on Earth.

When the United States began sending rockets out into space in 1958, one of the objects was to detect radiation beyond the atmosphere. Far more radiation was detected than had been expected.

The Earth, it seems, has a magnetic field, with a north magnetic pole in the Arctic, and a south magnetic pole in the Antarctic. Imaginary magnetic lines of force can be drawn from one magnetic pole to the other, each one representing a region of a certain fixed strength in the field surrounding the Earth.

A certain number of particles from the solar wind are trapped in the magnetic field. They tend to spiral tightly around the magnetic lines of force, moving back and forth between the magnetic poles.

Those regions surrounding the Earth (regions well beyond

the atmosphere) that are rich in trapped charged particles are called the "magnetosphere." To begin with, they were called "Van Allen Belts" because the American physicist James Alfred Van Allen was the head of the project that first detected and explained the radiation.

Not every planet has a magnetic field. Astronomers are not certain what it is that produces a magnetic field, but one attractive explanation involves a liquid central core capable of inducing a magnetic field if it is made to swirl, and a rate of planetary rotation that is fast enough to set up those swirls in the core. The Earth has a core of liquid iron, and rotates rapidly enough to produce the swirls, so it has a magnetic field.

Mars rotates with moderate rapidity, but has only a small iron core, if any. Venus must have an iron core, but it rotates very slowly. The moon lacks an iron core and rotates very slowly as well. All three bodies therefore lack a magnetic field, and lack a magnetosphere.

Jupiter, however, rotates very rapidly—more rapidly than any other body in the system. If it has some suitable core, it should have both a magnetic field and a magnetosphere.

As it happens, some of the radio waves arising from Jupiter can, in fact, be neatly explained by supposing them to originate from the energy lost by electrons as they spiral tightly about magnetic lines of force. Astronomers are therefore quite convinced that the magnetosphere of Jupiter has been detected, and shown to exist. They can even tell that Jupiter's magnetic poles are near, but not at, the geographic poles of the planet, something which is also true of Earth. Jupiter's magnetic poles are 8 degrees from its geographic poles, however, and are closer in angular measure than Earth's magnetic poles are to its geographic poles.

The radio waves of Jupiter emerge in bursts of energy at certain regular periods. In 1964 an American astronomer, E. R. Bigg, pointed out that these bursts seemed to coincide with certain positions of Io, the innermost of the Galilean satellites.

The exact reason for this is not yet known. The Galilean satellites exert tidal pulls on Jupiter, however, just as our moon does on Earth. Io, which is the closest of the Galileans, exerts the greatest tidal pull. This may contribute to the turbulence of the atmosphere in the direction of Io and account for the burst of radio waves.

The Inner Structure

How deep would the atmosphere be under Jupiter's visible cloud layer? The first guess might be just a few miles. On Earth, the clouds are anywhere from one to ten miles above its solid surface. The same is true for that other cloudy Terrestrial planet, Venus. Beams of radar waves have penetrated to the solid surface of Venus and have shown that it lies not far beneath the clouds.

But could this be true of Jupiter and the other Jovian planets? Here is where the question of their low density (see Table 5 on page 35) finally comes in.

Once it came to be realized that Jupiter had collected vast quantities of hydrogen, helium, and neon, it was no longer surprising that it and the other Jovian planets should be less dense than the Terrestrial planets. Hydrogen, helium, neon, ammonia, methane, are all much less dense than the rocky and metallic substances that make up the globe of the Earth and the other Terrestrial planets.

However, if we assume there is just enough of these low-density gaseous substances to make up a thin atmosphere around a rocky, metallic center, that would simply not be enough to account for the over-all low density of the Jovian planets.

The English astronomer Harold Jeffreys pointed this out in 1924.

He also considered the oblateness of Jupiter (see Table 17 on page 62). The amount of oblateness increases if there is a large difference in density between the outer regions and the inner regions of a planet. Suppose you begin with a planet that has Jupiter's mass and rotation period but is of uniform density all through. It will be oblate but not nearly as oblate as Jupiter is in reality.

If more and more mass is concentrated in the center, leaving the outer regions less and less dense, oblateness increases even though total mass and rate of rotation do not. The less dense outer regions are more easily lifted against gravity and are therefore lifted higher.

Jeffreys pointed out that between the low over-all density of Jupiter and the high concentration of mass in the deeper layers required to account for its oblateness, the planet would have to have an atmosphere some 4000 miles deep underneath the cloud layer.

Rupert Wildt took up this idea in 1938 (he had come to the United States to work by then) and went into more detail. It seemed to him that a 4000-mile-deep atmosphere would not remain gaseous. Under Jupiter's huge gravitational pull, the hydrogen-helium of the atmosphere would be compressed so tightly in the lower layers as to become solid even at temperatures far above those at which it would ordinarily melt and boil *under earthly conditions*. With this in mind, he began constructing what Jupiter's interior conditions might be like. . . .

Consider the Earth. The solid body of our planet is made up of a thick rocky layer about a metallic center. We see the rocky outside, of course, and we call it the "lithosphere" (from Greek words meaning "ball of rock").

Around the lithosphere is a rim of water, which is not quite deep enough to submerge all the land, but which permeates the land that sticks up above the ocean surface. This water layer is called the "hydrosphere" ("ball of water"). Finally, there are the gases above land and water which make up the "atmosphere" ("ball of vapor").

Earth's lithosphere is made up of complicated atoms that

cling together by strong electromagnetic forces. None of it would be lost in the course of the formation of the Earth. The hydrosphere and atmosphere, however, are made up of light molecules that do not cling together tightly, and are easily lost.

Jupiter, on the other hand, by holding on to everything, must have built up the hydrosphere and atmosphere enormously; not so much its lithosphere.

Wildt supposed, then, that Jupiter at its center would have a rocky and metallic lithosphere that would be small compared to the planet as a whole. Surrounding it would be an ocean of water, plus some ammonia, frozen rock-hard, of course, and this would be the hydrosphere. Surrounding that would be a thick layer of materials that on Earth would be gaseous and thin—hydrogen, helium, methane, ammonia vapor—but which on Jupiter would be compressed to material of such great density that it might as well be considered solid.

Wildt even calculated the size of each of these spheres, working it out so that he would end up with the right over-all density and the right concentration of mass at the center to account for the oblateness.

Suppose, then, one imagined oneself at the center of Jupiter, burrowing upward toward the surface. One would have to dig through a thickness of lithosphere, then of hydrosphere, then of atmosphere, in order, finally, to reach the clouds. How thick would each layer be? The result for each of the Jovian planets, according to the Wildt scheme, is given in Table 53.

As you see, Saturn, though smaller than Jupiter, is pictured with a much thicker atmosphere, one twice as deep as Jupiter. This would account for Saturn's unusually low over-all density and its unusually high oblateness.

From the standpoint of the Wildt structure, Earth is not quite such a pygmy as it seems, after all. It could be compared in volume not to the entire body of each Jovian planet, but only to the lithosphere, since Earth is virtually all lithosphere. In that case, we have the results shown in Table 54.

Even counting the lithospheres alone, every one of the Jovian planets is larger than Earth in terms of volume. The lithosphere of Jupiter is over a hundred times the volume of Earth, if this scheme of structure of the Jovian planets is correct. In terms of mass, the disparity is probably even greater, for under the huge weight of the hydrosphere and the atmosphere, the lithospheres of the Jovian planets are tightly packed and are denser than the Earth is.

TABLE 53

Possible Structure of the Jovian Planets

		JUPITER	SATURN	URANUS	NEPTUNE
Lithosphere radius	(in miles)	18,500	14,000	7,000	6,000
	(in kilometers)	30,000	22,500	11,000	9,700
Hydrosphere thickness	(in miles)	17,000	8,000	6,000	6,000
	(in kilometers)	27,000	13,000	9,700	9,700
Atmosphere thickness	(in miles)	8,000	16,000	3,000	2,000
	(in kilometers)	13,000	26,000	4,800	3,200

Nevertheless, the difference in volume or mass is not as great as it is when the Jovian planets are taken as a whole (see Tables 14 and 24 on pages 52 and 82).

The question is bound to arise, though, where all that rock and metal came from that serves as Jupiter's lithosphere? The lithosphere is built up out of material which makes up only 1 percent of the total matter of the universe. There was enough of the material at a distance of a hundred million miles from the sun to build up the Earth. It might be expected that there would be less material five hundred million miles from the sun.

Would there be enough out there to build up a lithosphere a hundred times the size of the Earth?

Why shouldn't Jupiter be mostly hydrogen and helium?

That seemed impossible, at first, since no matter how hydrogen and helium are compressed, their density remains (it was thought) insufficient to account for all the mass of Jupiter in the volume of the planet.

It turned out, however, that if hydrogen is placed under sufficient pressure, there comes a point where ordinary solid hydrogen suddenly collapses into a considerably denser solid, one in which the electrons have greater freedom to wander from atom to atom. This gives the new kind of solid hydrogen

TABLE 54

Volume of Planetary Lithospheres

PLANET	LITHOSPHERE VOLUME (VOLUME OF EARTH = 1)
Jupiter	104
Saturn	44
Uranus	5.5
Neptune	3.5

certain properties we associate with metals. We call this "metallic hydrogen," therefore, and find it can be considerably denser than ordinary solid hydrogen.

During the 1950s various astronomers tried to calculate a structure for Jupiter in which the center would be made up of metallic hydrogen and in which the end result would be to pack all the necessary mass into Jupiter's volume and to account for its oblateness, too.

In 1958 an American astronomer, W. C. DeMarcus, suggested that Jupiter be considered as made up of 78 percent hydrogen by mass, with almost all the rest helium. This means that there is 1 atom of helium for every 14 atoms of hydrogen, and gives Jupiter very much the chemical composition of the sun.

In such a planet, the pressure becomes great enough about 8000 miles below the cloud layer to turn the hydrogen, with its helium admixture, into metallic form. The metallic hydrogen would get denser and denser until by the time the center of Jupiter is reached the density of the hydrogen is 31 grams per cubic centimeter, which is 1.5 times as dense as platinum is on Earth.

There is no way, as yet, by which we can get direct evidence as to the inner structure of Jupiter and the other Jovian planets, but the DeMarcus theory seems to be most favored at the moment by astronomers.

Pioneer!

Naturally, the mysteries concerning Jupiter's atmosphere, its colors, the structure of its inner regions—and, of course, the nature, composition, and structure of its satellites and of surrounding space too—make astronomers long for a closer look.

On March 2, 1972, they set about getting just such a look, when a 570-pound (260-kilogram) Jupiter probe left Earth with an initial speed of 9 miles per second (14.5 kilometers per second), the fastest speed ever attained by a man-made object. This probe, appropriately called Pioneer 10, will move farther and deeper into space than anything sent up by mankind before.

It will make its way outward past Mars's orbit, into the asteroid belt. There it will send back information on the number and size of the fragments of matter it passes and, we hope, with which it does not collide.

Passing through the asteroid belt, it will eventually reach the vicinity of Jupiter on December 3, 1973, and will pass only 85,000 miles (135,000 kilometers) from Jupiter's surface, going right through the planet's magnetosphere.

During the four days it will take Pioneer 10 to fly by Jupiter, its 65 pounds (30 kilogram) of instruments, powered by four radioisotope batteries which, it is hoped, will survive the powerful radiation of the magnetosphere, will pick up radiation from Jupiter, count particles, measure magnetic fields, note temperatures, and analyze sunlight passing through Jupiter's atmosphere. At Pioneer's closest approach it will view Jupiter in the half-phase.

The data so gathered will go streaking back to Earth at the speed of light and, even so, will take 45 minutes to reach Earth. Even the path Pioneer 10 takes as it passes Jupiter will give us data: information on the mass of Jupiter and of its larger satellites. It should pass near at least one of the satellites and, if all goes well, fairly close to Io, Europa, and Ganymede, all three.

By the time it reaches Jupiter, Pioneer 10 will have lost much of its velocity, but in whipping about the giant planet it will regain speed. It will gain enough speed to break through the sun's gravitational grip and go skittering past the orbits of all the planets, noting, as it goes, how far outward the solar wind remains detectable. It may send messages back for three months after it passes Jupiter. In 1977, it will be passing the orbit of Saturn and by 1980 that of Uranus. In 1984, it will pass beyond Pluto. It will then be moving at a speed of 7 miles per second (11.5 kilometers per second) and will go out into the space between the stars, never to return to the solar system.

Pioneer 10 will be the first man-made object ever to leave the solar system, and it will take it some 100,000 years to cover a distance equal to that between ourselves and Alpha Centauri, the nearest star. Of course, it will not be moving in the direction of Alpha Centauri but, very likely, in the direction of the star Aldebaran. It would take Pioneer 10 about 1,700,000 years to reach the neighborhood of that star.

With Pioneer 10, as it goes out into the unknown, there will be a message from Earth, etched into a 6- by 9-inch, gold-covered aluminum slab. The message was designed by the American astronomers Frank Donald Drake and Carl Edward Sagan and drawn by Linda Sagan.

The most noticeable thing about the message is the outlined figure of a man and a woman, unclothed, and with the differences in sex indicated, thus giving certain minimum information about the kind of creatures that built Pioneer 10 and sent it on its way. The man is holding up his hand in what is hoped will be taken as a gesture of friendship and peace. If not, it at least shows the existence of four fingers and a thumb.

Behind the man and the woman is an outline of Pioneer 10 to scale. If beings find Pioneer 10 someday and measure its dimensions, they will know the size of human beings as well.

At the bottom of the slab are circles representing the sun and its nine planets, giving some indication of their relative sizes and of the rings about Saturn, together with a line marking the path of Pioneer 10 among the planets. That should be enough to identify the solar system as the place of origin of the probe.

Other symbols are included which express the location of the sun in the galaxy and which give an idea of our scientific advancement.

It might seem unwise to send a message out into the unknown like this. What if someone find it and decides to come to the solar system and conquer Earth?

In the first place, it is very unlikely that Pioneer 10 will ever come close enough to a star to pass through its planetary system. The chances are enormous that it will remain in deep space forever. Any intelligent beings who find it will have ways of traveling through deep space and, it is to be hoped, will be too advanced to feel any need to find a small planet, perhaps thousands of light-years away, and "conquer" it.

Secondly, we are in any case giving away the secret of our existence, and our location, by all the radio waves produced by our technological civilization, which streak into space in all directions at the speed of light.

No, if this message, the first ever sent out into deep space deliberately by mankind, is ever picked up, it will probably be millions of years hence when man himself may long have passed from the scene. There will then be at least one record that man existed and strove mightily to understand the universe about him.

Other Jupiters

Is Jupiter unique in the universe? Are there other Jupiters elsewhere, circling other stars?

Back in the 1930s, the answer might have been that probably few if any other Jupiters exist. In those days, astronomers had the idea that planets were formed only when two stars passed each other very closely so that the gravitational pull of each dragged matter out of the other. Out of this matter, the planets formed.

The stars are so far apart and move so slowly that the chance that any two might pass very close to each other is so minute that in all the history of the universe it is unlikely that a collision or near-collision would happen more than once or twice in each galaxy. Consequently, planetary systems would have to be very rare.

In the 1940s and afterward, however, additional evidence was gathered which made it seem that every star, as it forms out of a vast cloud of dust, would be accompanied by planets that form on the outskirts of the cloud.

If this is so, could we possibly detect the planets of other stars?

We can't hope to detect the planets by eye, of course. Even the brightest and nearest stars are mere sparks of light because of their vast distance from us. A planet near such a star, with a much smaller surface than the star, and shining only by such starlight as it could catch and reflect, would be so dim that it would be drowned out in the glare of the star it circled and not even our most powerful telescopes would have a chance to see it.

But we don't actually have to see an object to know it is there. What about the effect of its gravitational field? Earlier in the book, it was explained that Jupiter and the sun circle about their common center of gravity (see page 109). This means that the sun revolves about a point slightly outside its own sphere every twelve years—a wobble too small to detect at great distances, however.

But suppose the object that circled a star were more massive than Jupiter. Would it not force the star to make so large a circle about a center of gravity, that the wobble could be seen from Earth?

Such a wobble was detected as long ago as 1844, when the German astronomer Friedrich Wilhelm Bessel was studying the star Sirius, which is 8 light-years away. (This is not far for a star, but it is a vast distance of over 50 trillion miles.) Bessel detected a waviness in Sirius's motion and decided it was revolving about a center of gravity between itself and a companion. To account for the distance of the center of gravity from Sirius, the companion must have the mass of the sun. (In 1914 it was found that the companion, though it had the mass of the sun, was no larger than the planet Uranus. It was a "white dwarf.")

This is not exactly what we want, however. A companion as massive as the sun, more or less, can be detected. Jupiter, however, is only $\frac{1}{1000}$ the mass of the sun. Can an object as large as Jupiter be detected by a star's wobble?

So far, the answer is no, *if* the star it circles is as massive as the sun. But suppose the star it circles is considerably less massive than the sun. Then if a planet the size of Jupiter is moving around it, the center of gravity is farther from the small star than it would be from the sun, and it would wobble more noticeably.

In 1943 a Dutch-American astronomer, Peter Van de Kamp, was studying the motions of a star, 61 Cygni. Actually, the star included two stars, "61 Cygni A" and "61 Cygni B," which circled each other. From a distance it looked like a single speck of light, but a good telescope shows the two separate stars.

They are small stars, too, only about $\frac{1}{4}$ the mass of the sun, and Van de Kamp spotted one of them wobbling. He deduced the presence of a small third body, "61 Cygni C," which turned out to be only $\frac{1}{30}$ the mass of the star it circles and therefore about $\frac{1}{120}$ the mass of our sun. This makes it about 8 times the mass of Jupiter—a huge planet indeed, and the first to be discovered outside our own solar system.

Other huge planets have been discovered since. In 1963 the second-nearest star, "Barnard's star," only 6 light-years away, and only $\frac{1}{5}$ the mass of the sun, showed a wobble. In 1969, Van de Kamp calculated the wobble could best be explained by the presence of two planets, one about 1.1 times the mass of Jupiter and the other 0.8 times.

Both planets seem to be moving about Barnard's star in nearly circular orbits. The larger one is nearer Barnard's star, about the distance of the asteroid belt from our sun. The smaller is at a distance equal to that of Jupiter from our sun. The orbital periods are 12 years and 26 years respectively.

Altogether some six stars, each small and relatively near to us, have been found to have planets the size of Jupiter or greater. If so many Jupiters are found so close to us, when they are so hard to detect, we must conclude that Jupiters are very common objects and that a large percentage of stars are accompanied by Jupiters. This means that a large percentage of them have planetary systems, since they are unlikely to be accompanied by a Jupiter alone, and that there must be many Earths among the stars as well.

Yet though there may be many Jupiters, few of them may be exactly like our own Jupiter.

Consider Jupiter's center. Whether Jupiter is rocky and

Barnard's star is, in most ways, a thoroughly undistinguished star, small and ordinary. But it is the second-closest star to ourselves, and moves most rapidly across the sky of any star. It is the least fixed of the fixed stars. These photographs show the change in position of Barnard's star in merely 22 years.

August 24, 1894

May 30, 1916

metallic at the core or whether it is made up mostly of metallic hydrogen, the pressures at the center are enormous, thirty times the pressures at the center of our own small planet, Earth. The pressure at the center of Jupiter may be as large as 1,500,000,000 pounds per square inch, or 100,000,000 kilograms per square centimeter.

At such a pressure, atoms must be nearly at the point of collapse.

Under ordinary conditions, the tiny nuclei of atoms (which have only 1/100,000 the diameter of the atom as a whole) are surrounded by electrons. The electrons all have a negative electric charge and repel each other. The electrons on the outskirts of neighboring atoms, in repelling each other, prevent the atoms from coming too close together.

If, however, the pressure becomes great enough, the electrons are simply squeezed out of the way. The electron shell collapses and the tiny nuclei of the atoms are exposed. Neighboring nuclei can come much closer together than they could when they were protected by the electrons.

Almost all the mass of atoms is in the nuclei, and when the nuclei come close together they take up far less volume, but still have all the mass of the atoms. Such collapsed matter can become very dense, hundreds of times as dense as ordinary matter, many millions of times as dense in some cases.

Once atoms have collapsed and the nuclei can move about freely, they may smash into each other and undergo nuclear reactions. These nuclear reactions give off much more energy than do ordinary chemical reactions, which involve only the outer electrons of intact atoms. It is the nuclear reactions taking place in collapsed matter at the center of objects as large as the sun and other stars that keeps them radiating vast energies into space for billions of years.

Planets are cold bodies, precisely because the pressure at the center of such small bodies is not yet sufficiently great to collapse the atoms and set nuclear reactions going. But how big can a planet be before the collapse and the subsequent nuclear reactions?

It may be that Jupiter is just at that critical point. Recent measurements of the heat given off by Jupiter make that somewhat larger than the quantity of heat from the solar radiation falling upon it. Jupiter must have a source of heat of its own—perhaps some very occasional nuclear reactions taking place in its center.

Planets that are several times as massive as Jupiter, like 61 Cygni C, must have undergone sufficient collapse and enough nuclear reactions to grow rather warm, even at their surface. If the planet is large enough, say 50 times the mass of Jupiter, it may glow with a dim red heat.

It seems quite likely, then, that Jupiter is just about as large as a planet can be and still stay completely cold at the surface. Other stars may have larger bodies circling them, but they are no longer giant planets, but midget stars.

Our sun has the largest *true* planet possible—and its name is Jupiter.

GLOSSARY

ALBEDO—The fraction of sunlight reflected by a planet or satellite.

AMMONIA—A substance with molecules made up of a nitrogen atom and three hydrogen atoms (NH_3).

ANGLE—A figure formed when two straight lines meet at a point or when two planes meet along a line.

APHELION—The point in a planet's orbit where it is farthest from the sun.

AREA—The extent of the surface of any object.

ARGON—A gas making up $\frac{1}{100}$ of the Earth's atmosphere.

ASTEROID—A small planet.

ASTEROID BELT—The region between the orbits of Mars and Jupiter where most of the asteroids are to be found.

ASTRONOMICAL UNIT—The average distance of the Earth from the sun.

ATMOSPHERE—The layer of gases surrounding a planet, satellite, or star.

ATOM—A particle of matter made up of a central nucleus surrounded by electrons.

ATOMIC NUCLEUS (plural, atomic nuclei)—A tiny structure at the center of the atom, containing almost all the mass of the atom.

AXIAL TILT—The angle between the axis of rotation of an object and a line perpendicular to its plane of revolution.

AXIS OF ROTATION—The imaginary straight line about which an object spins.

CARBON DIOXIDE—A gas found in Earth's atmosphere in small quantities; found in the atmospheres of Mars and Venus in large quantities.

CENTER OF GRAVITY—The point about which two bodies move as each revolves about the other.

CENTIMETER—A measure of length, equal to about ⅖ of an inch.

CENTRIFUGAL EFFECT—The tendency for anything spinning about a center to move away from that center.

CIRCUMFERENCE—The total length of the curve of a circle or similar shape; the length of a curve marked around the widest part of a sphere or similar shape.

COMET—An object moving about the sun in an eccentric orbit and composed of materials which, on warming up near the sun, drift away from the comet to form a "tail."

DEGREE—An angular measure equal to $\frac{1}{360}$ the circumference of a circle.

DENSITY—The mass of an object divided by its volume.

DIAMETER—The length of a line passing across the widest part of a circle or sphere and through the center of the circle or sphere.

ECCENTRICITY—Degree to which the ellipse of an elliptical orbit is flattened.

ECLIPSE—The covering from view of an object in the sky by another object that moves in front of it.

ECLIPTIC—The plane that passes through the center of the sun and through all points of Earth's orbit.

ELECTROMAGNETIC FORCES—Electrical forces by which an electron is attracted to an atomic nucleus but is repelled by another electron.

ELECTRON—A tiny particle found in the outer region of an atom.

ELLIPSE—A curve that looks like a flattened circle.

EQUATOR—The circumference that lies halfway between the two poles of a spinning object.

EQUATORIAL BULGE—The extra thickness of a planet in the equatorial regions due to the centrifugal effect of its rotation.

EQUATORIAL DIAMETER—The diameter stretching from a point on the equator to the opposite point on the equator.

ESCAPE VELOCITY—The speed with which an object must move to escape from the gravitational pull of a planet, satellite, or star.

FOCUS (plural, foci)—One of two points inside an ellipse. The two foci are at equal distances from the center of the ellipse and on opposite sides, along the major axis.

FREE RADICAL—A molecule that does not have all the electrons between its atoms paired, but has one or more unpaired electrons.

GALILEAN SATELLITES—The four large satellites of Jupiter; so called because they were discovered by Galileo.

GRAM—A measure of weight equal to about ½₇ of an ounce.

GRAVITATION—The attraction exerted by one object on the other objects in the universe.

HELIUM—A gas composed of the second simplest of all atoms and found in Jupiter's atmosphere.

HYDROGEN—A gas composed of the simplest of all atoms and making up most of Jupiter's atmosphere.

HYDROGEN SULFIDE—A substance with molecules made up of two hydrogen atoms and a sulfur atom (H_2S).

HYDROSPHERE—The layer of liquid water on the surface of a planet.

INCLINATION—The angle between the plane of the orbit of a planet and the ecliptic.

JOVIAN PLANETS—The giant planets—Jupiter, Saturn, Uranus, and Neptune.

KILOGRAM—A measure of mass equal to 1000 grams, or about 2.2 pounds.

KILOMETER—A measure of length equal to about ⅝ of a mile.

KIRKWOOD'S GAPS—Regions in the asteroid belt where no asteroid orbits are to be found.

LIGHT-HOUR—The distance light travels in an hour.

LIGHT-MINUTE—The distance light travels in a minute.

LIGHT-SECOND—The distance light travels in a second; 186,282 miles.

LIGHT-YEAR—The distance light travels in a year; 5,880,000,-000,000 miles.

LITHOSPHERE—The rocky solid ball of a planet or satellite.

MAGNETIC FIELD—A region in which electromagnetic forces make themselves felt.

MAGNETIC LINES OF FORCE—Imaginary lines curving from one magnetic pole to another.

MAGNETIC POLES—Regions where the magnetic field is at its greatest strength.

MAGNETOSPHERE—A region outside a planet where particles from the solar wind are trapped along that planet's magnetic lines of force.

MAGNITUDE—A figure representing the apparent brightness of an object shining in the sky. The lower the figure, the brighter the object.

MAJOR AXIS—A diameter passing through the foci and center of an ellipse.

MASS—In a general way, the amount of matter in an object.

METHANE—A substance with molecules made up of a carbon atom and four hydrogen atoms (CH_4); found in Jupiter's atmosphere.

MINOR PLANET—An asteroid.

MINUTE OF ARC—An angular measure equal to $\frac{1}{60}$ of a degree.

MOLECULE—A group of atoms held together by electromagnetic forces.

NEON—A gas like helium but made up of more complicated atoms; found in Jupiter's atmosphere.

NITROGEN—A gas making up $\frac{4}{5}$ of the Earth's atmosphere.

NUCLEAR REACTIONS—Combinations between atomic nuclei that become possible when atoms have broken up at high temperatures and pressures, such as those at the center of stars.

OBLATE SPHEROID—An object with a shape that looks like a flattened sphere.

OBLATENESS—The flattening of a planet from spherical form because of the centrifugal force of rotation.

ORBIT—The path taken by an object revolving about a larger object.

ORBITAL SPEED—The speed with which an object moves in its orbit.

OXYGEN—An active gas making up ⅕ of the Earth's atmosphere.

PARALLAX—The apparent change of position of a close object compared to a more distant object, when the viewer shifts the position from which he views the object.

PERIHELION—The point in its orbit where a planet is closest to the sun.

PERIOD OF REVOLUTION—The time it takes an object to make one complete turn about a larger object.

PERIOD OF ROTATION—The time it takes an object to spin once on its axis.

PERTURBATION—A small gravitational effect, such as that of one planet on another.

PHASES—The different shapes of the lighted part of a planet or satellite that is shining by reflected light from the sun.

PLANE—A geometric figure that is perfectly flat and without thickness.

PLANET—Originally, an object in the sky which moved against the background of the stars. Now it is used for any object that circles a star, and shines only by reflected light.

PLANETOID—An asteroid.

POLAR DIAMETER—The diameter stretching from north pole to south pole.

POLES—The points where the axis of rotation reaches the surface of the rotating body.

PRIMARY—The planet about which a satellite revolves.

PROBE—A rocket-driven vessel designed to pass near some planet or satellite in order to gather information concerning it.

RADAR WAVES—Like light waves, but much longer (and therefore invisible).

RADIO WAVES—Like radar waves, but still longer.

RETROGRADE MOTION—Motion in a direction opposite to what is usual.

REVOLUTION—The circling of an object about another object.

ROTATION—The spinning of an object about its own central axis.

SATELLITE—An object that revolves about a planet.

SECOND OF ARC—An angular measure equal to $\frac{1}{60}$ of a minute of arc.

SHORT-PERIOD COMET—A comet whose period of revolution is 100 years or less.

SOLAR SYSTEM—The sun and all the objects that are held in its gravitational field and that move around it.

SOLAR WIND—Charged particles from the sun moving out at high velocity in every direction.

SPECTRUM (plural, spectra)—Light which has been spread out so that each different wavelength is in a different position, as in a rainbow.

STAR—One of the objects visible in the sky as a tiny spark of light.

SULFUR—A solid, yellow substance, possibly found in Jupiter's atmosphere.

SURFACE—The outside of any solid object.

SURFACE GRAVITY—The strength of the gravitational pull on the surface of a planet or satellite.

TELESCOPE—A tube, containing lenses or mirrors, which makes distant objects look larger, nearer, and brighter.

TERRESTRIAL PLANETS—The small planets—Mercury, Venus, Earth, Mars, and Pluto.

TROJAN ASTEROIDS—Asteroids traveling in Jupiter's orbit, but either 60° ahead or 60° behind Jupiter.

VAN ALLEN BELTS—The magnetosphere.

VOLUME—The room taken up by any object.

WATER—A substance with molecules made up of two hydrogen atoms and an oxygen atom (H_2O); common on both Earth and Jupiter.

WHITE DWARF—A pygmy star, no larger than a planet, but just as hot and massive as an ordinary star.

ZENITH—The point in the sky that is directly overhead.

INDEX

Italics indicate illustration